# Creating the Human Past:

## An Epistemology of Pleistocene Archaeology

Robert G. Bednarik

Archaeopress

Gordon House
276 Banbury Road
Oxford OX2 7ED

www.archaeopress.com

ISBN 978 1 905739 63 9



on the cover: Pleistocene excavation in the massive Wonderwerk Cave, South Africa. The sediment beneath the plastic bucket contains Oldowan stone tools and the most ancient hearth known in the world, 1.7 million years old (photograph by Peter B. Beaumont).



Printed in England by Information Press, Oxford

# CONTENTS

# 1

# INTRODUCTION
## *Preamble*

This book has been overdue for at least a century. Archaeology has operated for well over 150 years as a politically and ideologically influential discipline, but in all that time it has not been severely taken to task over its systematic mistakes, the haphazard way it forms its notions about the human past, or many other relevant aspects of its operation as an academic pursuit. It is essential, for its continued survival, and as a prelude to its inevitable renewal, to examine the epistemological foundation of archaeology, and to consider its development over time.

Archaeology is usually defined as the study of the past through the *systematic recovery and analysis of 'material culture'* (e.g. in Paul Bahn's *Collins Dictionary of Archaeology*). Its primary aim is to *recover, describe and classify* material remains considered to be of archaeological relevance, and from this the form and behaviour of past societies are then deduced. In a superficial way this definition may sound convincing enough, but when we begin to look at it more closely, questions soon arise.

What does this term 'material culture', which we see so often used in archaeology, actually mean? It is clear that it refers to kinds of objects archaeologists recover from excavations or observe elsewhere in the landscape, which refer in some way to past cultures. But are these cultural remains *representative* of the societies who produced or used them? Of course not, most cultural material of the past left no trace at all, for example song, dance, mime, language, mythology and so on. Where material traces of cultures actually did survive, they are in most cases mere shadows of what may have existed once. There are very few exceptions to this rule, such as stone implements, which have a comparatively high rate of survival. This is particularly relevant when we consider the Pleistocene period (the Ice Ages), which accounts for most of human history, and which can be assumed to have been subjected to much more taphonomic distortion than the Holocene, i.e. the last 10,500 years.

The conjunction of the words 'analysis' and 'to classify' as used in the above definition of archaeology calls itself for analysis. Most of this classification is quite subjective, because most collections of entities (e.g. artefacts, structures) do not present us with apparently solid bases for categorisation: they offer us no periodic tables of elements; they do not even consist of species. Most taxonomies, even in science (including in biology), are in a state of flux, being contingent upon historical developments in the discipline concerned. In the sciences they do have one redeeming feature, however: they are falsifiable, they can be tested through processes of refutation. This is not the case with the interpretations archaeology can offer us, where non-inductive experiments are not possible. So it must be stated quite categorically that the classifications archaeology produces should be considered *arbitrary constructs of specialists*. Perhaps they are valid, perhaps *some* of them are — or perhaps *none*. We cannot readily test taxonomic propositions in archaeology.

Even less can we 'analyse' them. In scientific usage, 'analysis' refers to a separation of an entity into its components, and to a rigorous examination of its constituent elements. But as an eminent South African archaeologist, Professor Lewis-Williams (1993), has pointed out, these 'elements' archaeologists might perceive in remains of 'material culture' are creations of the researchers themselves. In reviewing the destructive activity of the discipline, Australian archaeologist David Frankel (1993) has defined the work of his peers as being similar to that of the sculptor. The individual archaeologist 'finds' interpretation just like the sculptor 'discovers' a statue in a block of marble. In both cases, the interpretations are products of creativity, and naturally they will differ between practitioners. Moreover, these 'egofacts', as Uruguayan archaeologist Mario Consens (2006) calls them, should be expected to differ according to the historical context in which they are offered for our consideration. The practitioner does not exist in a cultural vacuum at any one juncture in history, and even less in an academically neutral state. On the contrary, there are many currents that determine what interpretations are preferred in shaping what British archaeologist Paul Bahn (1990) has called the 'accepted fiction' archaeologists favour at any given time. Foremost among them are the powerful dogmas this discipline has developed. To then 'analyse' these taxonomies of archaeology's consensus models tells us about how devout archaeologists perceive aspects of the physical world, about the preoccupations of archaeologists, and about their contingent prejudices. It cannot possibly tell us anything *reliable* about the human past. Certainly, some of archaeology's interpretations are likely to be valid, perhaps even many of them, but without the facility of testing them we cannot expect ever to know which ones, or what proportion of them, we can trust.

A scientific analysis of archaeology's favoured model of the past, as well as of its nomenclatures, at any one point in history is therefore capable of telling us a great deal about the discipline itself, its academic and heuristic dynamics, its politics, its evolution through history (Habermas 1979). It does not, however, tell us anything about the subject of archaeology, the peoples of the distant past, with any semblance of scientific rigour. In a general sense, archaeology is an academic pursuit whose role it is to create for contemporary societies the modern myths about the distant human past. Because it avails itself of a great variety of scientific procedures in this quest, many of its interpretations are likely to be 'true', or at least partially valid. The importation of falsifiable propositions from scientific disciplines does not, however, automatically confer a scientific status on archaeology itself. The conditions for this would be considerably more demanding. Other human pursuits, such as industry or technology, also import scientific knowledge claims, but that does not make them sciences.

To illustrate one of the differences between archaeology and scientific pursuits, let us consider the following example. There are a number of disciplines that deal with events and phenomena of the past (for instance geology, palaeontology, sedimentology). Some of these, when they are conducted within certain rules, are scientific, others are not, even though they may be based on perhaps perfectly 'sound' practices. Consider the similarities and differences between astronomy and archaeology. Both deal with the past; no astronomer has ever observed an event or phenomenon of the present (for the sake of the argument, we shall ignore here the question of linear versus non-linear time). He or she can only witness the *past* in cosmic space, because cosmic present is only rendered accessible to us by *becoming* cosmic past. Some of the astronomical events we observe occurred some minutes before certain of their effects become detectible to us, others took place many millions of years ago. But despite the similarity of dealing with events and phenomena of the past, there are significant differences between astronomy and archaeology. The astronomer can make predictions about the trajectories of all sorts of variables and then test them; the archaeologist cannot. The astronomer uses universals from physics in explaining observations (e.g. spectral shift, properties of chemical elements, nuclear reactions), whereas those cited by the archaeologist refer to ethnographic analogy, deductive uniformitarianism or similarities in the products of modern experimentation (e.g. microwear on implements). Many of these explanations may be valid, perhaps even most of them. This is not the issue; the issue is that there is no mechanism available to us to test them.

These considerations, one would assume, might prompt archaeology to be open-minded, receptive to criticism and to alternative paradigms. Many individual practitioners certainly are, but the discipline as a whole is, as we shall see later in this book, hostile to challenges of its dogmas. In my experience its intransigence is not greatly different from that found in other belief systems, such as religions. The actual merits of an argument or of the evidence in question are of little concern once archaeological debates become imbued by ethnocentrism, nationalism, jingoism, academic sectarianism, or by a desire to preserve a status quo, to crush academic dissent, or to preclude interlopers from other disciplines from swaying archaeological thought. When we consider that these tendencies happen to coincide with the non-refutable character of many, if not most, archaeological interpretations, it becomes apparent that such a combination would tend to restrict the discipline's ability to exercise self-criticism. It must be expected to lead to a 'sluggish' discipline, one that discourages innovativeness and resents scrutiny of its dogmas. It is likely to regard meddling 'outsiders' or epistemological 'renegades' with suspicion or respond with hostility, particularly when these seem to challenge the established authorities in the discipline. In such an academic climate, models most likely to flourish will be those that are most compatible with mainstream ideology and are least refutable. This creates an unhealthy epistemological climate for the discipline, favouring non-scientific directions and academic partisanship. The historical implications of this will be illustrated in this book with a number of cases studies.

Academic practices have a tendency to trap researchers in their own creations even at the best of times, because they encourage selective acquisition of confirming 'evidence' and specious defence of favoured models. In academia, there are no points to be scored for falsifying one's own theories, or for readily conceding that someone else had falsified them. The competitive academic system that has evolved, especially in the Western world, encourages the individual implicitly to defend her or his hypothesis at all cost. Being shown to be wrong or admitting to being wrong is regarded as weakening one's professional standing. Academic rewards are restricted to those who prevail, the verbally facile, the unyielding — and those who are careful enough to couch their claims in non-refutable terms. All of this runs counter to scientific ideals, it is rather more reminiscent of religious fundamentalism. The true scientist lacks all certainty, just as the mark of the real scholar is a profound form of academic humility: he (or she) does not know whether he is right, he does not even expect to ever find out. True science acknowledges that humans have no access to 'objective reality'. So there are no absolutes in human knowledge, which is in a constant state of flux, based as it is on the rather modest intellectual and cognitive means our evolution has equipped us with.

As one of the 'social sciences' (almost an oxymoron, as the ability of an organism to study itself objectively at the unsophisticated level these disciplines operate must be questioned; Bednarik 2011), archaeology is subject to certain obvious limitations. By its very nature, theory has to abstract features, attributes, factors, etc. from their pragmatic context, and relate these elements by abstract laws or rules. This strategy works in the 'hard sciences', but in the 'social sciences', where what counts as the facts in a given situation depends on contextual interpretation, the attempt to decontextualise the elements over which theory ranges can only result in approximate predictions. The social 'sciences' cannot at the best of times achieve the definitive predictive success that underlies the disciplines of hard science, nor can consilience be found with them. The obvious solution for the social 'sciences' is to treat the background skills used in everyday contextual interpretation as a formalisable *belief system*, and thus integrate contextual interpretation into their theory. Moreover, even among the 'social sciences', archaeology is incompatible with the rest of them, because the methods of data gathering available to those others are simply not there to archaeology. The societies in question do not exist, and over 99% of them have left no explanatory records whatsoever.

Another factor to be considered in evaluating archaeology is that it is a fundamentally destructive pursuit: its principal tool of enquiry, which dominates its methodology, is excavation. The British pioneer archaeologist Sir Mortimer Wheeler compared excavation of soil-like sediments for the purpose of finding selected types of material remains in them to reading a book whose pages become blank as soon as they have been read the first time. This simile can be extended further. There should be no doubt that even the most accomplished archaeologist would only be able to 'read' a very few of the words in this book before the text disappears. The types of information he or she looks for would be only a small sample in the vast spectrum of possibilities, and this sample would be determined by the knowledge, preoccupations and skill available to the excavator, among other factors. He or she would only look for evidence of certain types, and not for other types. In practice, the excavator's priorities will be conditioned by such factors as available analytical technology, research project design, the time and labour available, the available funding, academic conditioning of researcher and referees, and a variety of others — but most importantly, by the limitations of knowledge of the excavator. The actual excavation will in most cases not even be done by the experienced project director. Most archaeological excavation is in fact done by students, volunteers and paid labourers.

Not only is all excavation destructive, there are other, less obvious factors involved. For instance, Egyptologist John Romer has documented examples of recklessness in contemporary research. In the Valley of the Kings, at Thebes, he has shown that archaeological work has been destructive in unexpected ways. The numerous tombs there are hewn into limestone that rests on a hygroscopic shale facies. By opening the tombs and excavating them on a large scale, increased evaporation of moisture from the shale has led to shrinkage and geophysical adjustment, which caused stress fractures in the limestone, and rampant damage of the tomb walls and roofs. The tombs are now themselves crumbling because the natural equilibrium has been disturbed by many decades of archaeological activity. When requested to make provision for proper conservation treatment in their projects, Egyptologists point out that they have no experience in structural conservation measures; that these are costly and that research sponsors cannot afford to underwrite the substantial costs of preserving structures *in situ*. Yet these structures had previously been preserved perfectly for millennia. It has been argued that the looting by 'professionals', which was begun in Egypt in the early nineteenth century, is still going on there, now under the guise of archaeology. Egyptology has long ceased to produce new knowledge of great importance; it has become a routine industry. Practitioners are more concerned, some say, about preserving Pharaonic culture in more obscure tomes, in writing their papers and theses and in climbing the academic ladder of the discipline, than in preserving this heritage for future generations.

This is of course just an example of a much deeper malaise. There are many ways in which archaeology endangers and destroys archaeological resources. Rock art, for instance, has on countless occasions been recorded by destructive methods, or has been destroyed, or allowed to be destroyed, by the very same archaeologists who were placed in charge of its protection. There are examples of this from throughout the world, ranging from the sawing off of whole panels, from Karelia to Australia, to the use of inappropriate contact recording methods. Archaeology has looted and stolen millions of items of 'material culture' from their native regions, ranging from the Elgin marbles Britain stole from Greece to the human body parts scavenged from graves in Tasmania, to Priam's golden hoard from Troy, smuggled out of Turkey, held in Berlin and seized by the Red Army (Simpson 1997). Napoleon looted much of Europe and North Africa to prove France the Roman Empire's rightful heir. Nationalistic chauvinism underpins the rapine of objects like the Rosetta Stone, 'honourably acquired by the fortune of war' and now held by the British Museum (consider, for instance, its refusal to return the remains of *Proconsul africanus* to Kenya). Archaeology sheds crocodile tears over the looting of archaeological resources by the suppliers of the illicit antiquities trade, while ignoring that these materials only became commodities through the promotion of archaeology (Elia 1996). There are indeed numerous facets to just the issue of equating heritage with identity. On the one hand, heritage is proclaimed to be the legacy of all humanity; on the other it is the hostage of nationalism. It stands to reason that archaeology, with its penchant for public support, needs to be examined critically.

It needs to be emphasised from the outset that many archaeologists fully recognise weaknesses of archaeology, and have often looked for ways of alleviating them. A recent example is the 'Campaign for Sensible Archaeology' *(http://www.facebook.com/group.php?gid=123023784380067)* which raises three principal criticisms: that the language of archaeology is "unpleasantly obtuse and dense"; "the disregard of factual evidence in favour of opinions and speculation"; and "deliberately stretching the boundaries of what is considered suitable for archaeological study, particularly projects investigating material from the very recent past and even into the present". The second and third concerns will be addressed repeatedly throughout this volume.

## Generic problems

Rather than belabouring specific problems such as those canvassed above I wish to focus on the field's generic quandaries. Archaeology as a discipline possesses no autonomous universal theory. Its theoretical underpinnings are a potpourri of theories and scraps of theories, imported, often in corrupted form, from other disciplines. Uniformitarianism has served geology and other fields well, so a particular brand of it, modulated by selective ethnographic analogy, provides the discipline's de facto universal theory. It facilitates the view of past human societies as mechanistic entities, in the same determinist way one would study other organisms. But humans have always been 'intelligent' organisms with highly complex cultural imperatives, throughout their history, and one must question the adequacy of this approach. Human responses were no doubt always influenced by cultural choices, by decisions that bore little or no resemblance to the action-response models prescribed by determinism. There is no allowance for individual initiative in processual archaeology (see Chapter 2), in fact this form of theory effectively reduces its subjects to organisms of predictable behaviour patterns that played out their roles in 'prehistory' in the same uniformitarian way sand grains being washed down a slope behave entirely as one could predict.

A major misunderstanding about archaeology is the belief that there exists some homogeneous entity called 'world archaeology'. This is a myth. The concept of archaeology has quite different meanings in different parts of the world, and these may be determined by political, ethnic, cultural and religious preoccupations of societies. In the U.S.A., archaeology is a sub-discipline of anthropology, whereas in many other world regions it is an autonomous discipline, a collection of quite diverse concerns ranging from numismatics to Pliocene hominoid evolution. As archaeologists Philip L. Kohl and Clare Fawcett (2000: 13) observe, "Most of the recognised 'regional traditions' of archaeological research are in fact *national* traditions which have developed within the framework of specific nation-states". The politically determined diverse spheres of interest seem to be held together particularly by the method of excavation. But this is not a technique of investigation exclusive to archaeology; it is shared with many other disciplines, such as palaeontology, sedimentology, palynology and geology. In various schools of archaeology, the term 'prehistory' is preferred, which only serves to illustrate the ethnocentrism of this discipline. Based historically on antiquarianism and the pursuit of ethnic and religious origins, this form of archaeology ignores that the term 'prehistory' is likely to be offensive to more than 90% of all humans and human societies that ever existed. The term is itself unscientific, because the implied proposition concerning the significance of written records (that they are more reliable than oral records) is unfalsifiable. The introduction of writing does have huge scientific consequences, particularly in the neurosciences (Bednarik 2012), but these are totally different from the simplistic understanding the use of the term 'prehistoric' implies.

In addition to archaeology's lack of falsifiability, which bars it from scientific status, there are other reasons precluding such a position. Among them are the controversies over the curatorial ambitions characterising the discipline. It often seeks control of access to data, objects, sites and so forth, which has led to confrontations particularly with indigenous peoples (e.g. over the possession of skeletal remains or particular artefacts, or over the dissemination of certain restricted knowledge). This raises the issue of archaeology's political roles. The discipline arose largely from the need to underpin the emerging nation-states in the 19th century, imbuing them with early histories and origins myths. Since then the states have gained complete control of the discipline — training and licensing all archaeologists and employing nearly all. This means that in a country such as Australia, where most archaeology refers to the history of the indigenes, the state exercises control over all archaeological sites, finds and data. Bearing in mind that the militarily defeated or colonised autochthons have no reason to like or to recognise the states that usurped their sovereignty, this is then a case of adding insult to injury. Politically they object to the archaeology of the occupying power as just another form of colonialism, cognitive colonialism, and there have been heated battles between local indigenes and archaeologists in various parts of the recently colonised world.

Much ink has been spilt over the political roles of archaeology, and yet there are many professional archaeologists who still reject that archaeology has a political role. But all over the world, it is archaeologists who manage the remains and monuments of the defeated, marginalised and superseded cultures for the victorious states whose servants they are. It is the archaeologist who decides whether there was a previous Hindu or Jewish temple at the site where a mosque now stands (a decision likely to involve much bloodshed), and it is the archaeologist who decides by what means the victims of this or that mass grave met their end. Throughout the history of the discipline, archaeologists have created fictitious grandiose pasts for nation states, most especially in dictatorships. Examples can be cited from all over the world, but most especially from Europe. Just as the archaeologists of the former Soviet Union were obliged to serve their political masters, many of the fierce nationalist movements in modern Russia are led by archaeologists and historians. As the historian E. J. Hobsbawm (1992: 3) stated, "historians are to nationalism what poppy growers in Pakistan are to heroin addicts; we supply the essential raw material for the market". To which the archaeologists Kohl and Fawcett (2000: 13) added, "rather than just the producers of raw materials, historians and archaeologists may occasionally resemble more the pushers of these mind-bending substances on urban streets, if not the mob capos running all stages of the sordid operation". The political uses made of archaeology's "findings have facilitated ethnic clashes and cleansing, bigotry and nationalism far more often than they have promoted social justice" (ibid.). Such comments are perhaps primarily intended to refer to the involvement of archaeologists in the USSR, Nazi Germany, Salazar's Portugal, Franco's Spain, to the Balkan countries and their archaeologically supported rampant nationalism, as well as that of the Caucasus region or the Near East or apartheid South Africa, among others — but even in the most 'democratic' countries, archaeology can have sinister overtones. For instance most Australian archaeologists would scoff at the suggestion that they have political roles, but they do. One of countless examples illustrating the point is the plight of the rock art precinct of the Dampier Archipelago, on the continent's northwestern coast. It is regarded as the world's largest concentration of petroglyphs, and its destruction by industry since the 1960s has been greatly facilitated by archaeologists, particularly since about 1980 when archaeologists began supervising the controlled destruction of countless rock art sites. So much so that when I launched a major campaign to save this incredible monument I found, to my amazement, almost no support of it among Australian archaeologists. Their argument, no doubt, was that they should not be seen as politically active, when in fact they were more concerned about what would happen to their lucrative consultancy contracts with the immensely powerful corporate interests operating at the site. Many similar examples can be cited from throughout the world.

The notion of idealism and the political neutrality of archaeology derives very little support from reality. Archaeologist Neil Asher Silberman (2000), in a paper entitled 'Promised lands and chosen peoples: the politics and poetics of archaeological narrative', speaks of "the archaeologist with a thousand faces", and especially the "Archaeologist as Hero" (the John Cullinane and Indiana Jones figures we are well familiar with). Bruce Trigger (1984, 1989), yet another archaeologist, divides archaeologies into nationalist, colonialist and imperialist, to which Silberman adds two more categories, touristic archaeology and an 'archaeology of protest'. Archaeologists who refuse to accept that their discipline is politically active have apparently never given any thought to the matter.

Another aspect of their discipline needing attention is its vexatious relationship with religion, which we will return to later. The most obvious manifestation of this is Biblical Archaeology, a field where religious preoccupations are frequently so intertwined with the pretence of an academic pursuit that its value to learning is hardly self-evident. However, there are many less obvious correlations with religion. It is not at all surprising that many of the greatest 'prehistorians' were men of the cloth, particularly for the century after Darwin's *Origin of the species* in 1859. Once the Church realised the threat of evolutionist ideology it sought to inform itself through encouraging the pursuit of archaeology by its priests. This had the added benefit of watering down the more strident strains of fervency in the discipline. Many aspects of it soon reflected a mild theocracy, for instance the way Palaeolithic cave art 'sanctuaries' (note the terminology used) were validated resembled the way religious shrines were (Freeman 1994). Still today we have a Biblical terminology to define supposedly secular archaeological concepts, such as the 'African Eve' or 'African Adam', or the 'Garden of Eden'. Still today archaeology operates on the basis of confirmation (seeking to confirm that which is already assumed to be true), the framework that sustains religions but which is the very opposite of refutation, the way of science. And still today devout archaeologists are apprehensive of science, fearing its methodology and occasionally attacking its practitioners when they turn their attention to archaeology. Indeed, recently an archaeologist, universally agreed to be one of the finest America has produced, published a paper entitled "On science bashing: a bashful archaeologist speaks out", in which the founder of what has come to be called the New Archaeology said:

> "Humanists [in archaeology] are committed to the defense of their chosen identity. Their methods are vacuous and their attempts at learning pathetic. When challenged, their only recourse is to *ad hominem* argument. Those who do not share their privileged knowledge are to be understood as defective persons, persons blinded to the truth, or persons who deny the truth in order to pursue dubious social goals." (Binford 2000–2001: 334)

Humanist archaeology's fear of science seems entirely irrational, because practically all archaeological progress nowadays is provided by the sciences, especially physics, chemistry and the earth sciences. Thus on the one hand, scientific data and propositions are eagerly imported from the sciences, but on the other hand the methodology of science is categorically rejected in favour of the discipline's de-facto universal theory of latent uniformitarianism and ethnographic analogy. To cushion archaeology from the 'harshness' of science, a field called archaeometry has been created some decades ago. It seems to be intended as a kind of hybrid discipline, but, having attended many such conferences, it seems to me more like a refuge, a patch of neutral turf where the two philosophically incompatible sides meet ritually.

Another generic problem with archaeology concerns matters that I am a little reluctant to raise, because I know from experience that overzealous archaeologists tend to become agitated when I do. But in the interest of explaining generic problems with the discipline I have to find a way of conveying this here, and do so as gently as possible. Some archaeologists are exceptionally well informed and competent, but many have surprisingly low standards of archaeological knowledge. At least in part this is related to the fragmentation of the discipline into regional and usually national 'schools', and the lack of effective dialogue between these. The Anglo-American school, for instance, seems to assume that everything of any consequence has been published in English. Not only has more than 80% of all archaeological knowledge never been made available in that language, much of what has been published has appeared in exotic, unknown journals or volumes. So here the problem is one of academic parochialism. By contrast, I have never met a Russian archaeologist who is not fully fluent in at least two languages, but most seem to manage several. Between them, the scholars at a major Russian archaeology department can probably read most languages archaeological material has appeared in. Much the same applies in many other parts of the world. This tends to yield a bland conformist version of the discipline that is unaware of its limitations. To illustrate with an example: the knowledge that *Homo erectus* managed to colonise island Wallacea has been available for almost five decades, but had not been published in English until recently. Much the same applies to most of the information concerning early art beginnings, and numerous other special fields or methods (examples will crop up later in this volume). The problem with such a profound lack of archaeological knowledge at many Anglophone university departments is of course that it limits the information available to their students, it encourages more parochialism in the next generation, and it renders the constructs of the human past promoted by these institutions hopelessly skewed.

I apologise to any scholar who feels offended by my bluntness, but this is an important point to make here. This book is a sincere attempt to address legitimate concerns; it is hopefully free of gratuitous critique, and it is intended to communicate, to facilitate improvement of the discipline. A somewhat dismal picture of archaeology would emerge if this attempt to deconstruct it were to lead to the view that the discipline is incapable of learning from its mistakes in the way it has historically treated dissident scholars (see Chapter 4).

## When archaeology turns feral

Most archaeologists of the world work, directly or indirectly (as consultants), for the state, and their discipline is an institution of the state. Yet from the perspective of the people of long-gone cultures, these states usurp their histories. There are very few states in the world today whose sovereignty was not acquired through war, conquest, genocide, violent colonialism or atrocious suppression of previous societies. Just as all Histories is inevitably written by the winners, most pre-Histories deal with the losers, the societies supplanted or extinguished. The study of these 'loser societies' by the state that represents the usurping 'winners' will always be a political process. If it is conducted by agents of today's state it is a re-writing of history by our contemporary governments. Some archaeologists will scoff at this truism, which already indicates how biased their judgment is, and how inadequately they are qualified to objectively and sensitively interpret the history of previous peoples. Other archaeologists do accept its validity, but argue that some forms of archaeology do make an effort to overcome the fact that their practitioners serve political masters.

While no doubt correct, it is also true that both archaeological and anthropological research have been used to support the hegemony of imperialist powers, as well as the subjection of indigenous peoples. The discipline was formed during the 19th century in response to nationalistic needs, as already mentioned. The inherent ambiguity of all archaeological data lends itself ideally to the hegemonic interpretation of the past in terms of current political concerns. Any careful study of the history of archaeology during the last two centuries will reveal that archaeological 'interpretations' and even priorities merely reflect contingent politics of the time in question.

Enlightened archaeologists have suggested that since archaeological interpretation is a form of political discourse it should be subject to the same standards of public accountability as other forms of expression (e.g. Silberman 2000: 250).

The political dimension is not limited to archaeology, it has also been encountered in a variety of contexts in anthropology. Social scientists, especially psychologists and anthropologists, have for many decades been engaged in such areas as interrogation techniques, counter-insurgency policies, methods of torture and intelligence gathering, and other areas of partisan use of the social and behavioural sciences (cf. Escobar 1991: 659; Price 2000, 2005; Houtman 2006, 2007; McNamara 2007). For instance, there has been much debate about the recruitment of anthropologists as spies, e.g. by the Central Intelligence Agency of the United States (CIA). The Pat Roberts Intelligence Scholars Program (PRISP) or the Intelligence Community Scholars Program (ICSP) provide other examples of pathological anthropology in the United States, while corresponding programs in Canada or Australia are perhaps more secretive or more subtle. In the U.S.A., jobs for anthropologists to work for the CIA have been openly advertised through such bodies as the American Anthropological Association. Such academics are required to provide briefings "directly to senior policy-makers and military commands". Covert researchers are encouraged to attend academic conferences, where they must "show a high tolerance for ambiguity" — whatever that might mean. Such anthropologists will also have a high tolerance for the CIA's long history of torture, terrorism and covert support for anti-democratic movements anywhere in the world. The incursion of the CIA into the discipline of anthropology is well illustrated by the removal, in 1990, of prohibitions against covert research in the AAA's *Principles of Professional Responsibility*, its code of ethics.

The concept of a 'pathological archaeology', on the other hand, has not been much discussed so far (but see Bednarik 2006, 2007, 2008). Since the time Australian indigenes gained a political voice, after it was decreed by referendum in 1967 that they be counted as people (removing Section 127 of the constitution), and their prior settlement of the continent was legally acknowledged in 1992, they have often expressed their opposition to archaeological and anthropological practices. Even in recent years, archaeology professors still fought Aborigines in the courts over custodianship of archaeological materials. Skeletal remains arrive in Australia every year from museums abroad, having been supplied by the grave robbers of earlier times. Some Australian archaeologists still exist in the delusional state of believing that they represent science and therefore have inalienable academic rights that should have precedence over indigenous rights. But we have already seen that archaeology as currently practiced by the state is not a science; it is a political pursuit of interpreting the human past from a biased perspective. Moreover, science has no custodial demands and it has no agenda of academic exclusion — as state archaeology certainly does.

The purest expression of a pathological archaeology, however, is the participation of archaeologists in the deliberate, systematic and needless destruction of archaeological monuments, such as rock art sites or stone arrangements in remote regions (Dyson 1997; Arcà et al. 2001; Bustamente 2006; Bednarik 2007). For instance, many millions of dollars have been paid to archaeologists at Dampier Archipelago, Western Australia, to facilitate the perverse destruction of the world's largest concentration of rock art. The objections of the owners of the monument, the local indigenes, were ignored in this. No use was made of the protective legislation of Western Australia concerning the rock art, and when the responsible public authorities were challenged by concerned outsiders to exercise their responsibilities, they failed to do so. The underlying issue is succinctly expressed by the late Vine Deloria, a First Nations leader in the U.S.A.: "Western civilization, unfortunately, does not link knowledge and morality but rather, it connects knowledge and power and makes them equivalent."

## The unsatisfactory state of the discipline

Another summary view of archaeology was bluntly expressed by an influential Australian writer, Frank Campbell (2006):

> "Archaeologists dig up their own future. And there's the rub: their careers depend on what they find, how important their finds and how others interpret them. Careers are at stake. There are very few decent jobs. There's a nasty hierarchy to negotiate. ... Archaeologists dig up someone else's past, which means nothing but trouble. ... From Wales to Australia to Jordan, the present molests the past for its own nefarious purposes. ... If careerism and nationalism were all archaeologists had to worry about, they'd be laughing and drinking instead of just drinking. The tragedy is that archaeology has promised a grand narrative but can deliver only conjecture. The archaeologist has no clothes."

If this were a preview of the direction into which public perception might be developing, it would not augur well for the discipline's future. In contrast to other fields of academia, archaeology produces nothing of economic value (unless the production of TV films is considered to be of economic value). It therefore depends much on the public's favour, or indeed, its benevolence. If society at large were to discover that the greatest threat to cultural heritage does not come from tourists, looters or amateur archaeologists, but from professional practitioners, it might well become inclined to withdraw its patronage from public archaeology. In recent years much effort to enthuse the public's interest in archaeology has become evident, especially through a variety of television programs, ranging from hard documentary to reality shows and imaginative interpretation. An excellent vehicle of public education, such programs tend to portray archaeology in the most positive terms, and I have been involved in the production of many of them. However, much of this rapport with an admiring public depends on maintaining the image of the archaeologist as the intrepid truth-seeker, a font of archaeological wisdom, a fine 'scientist' working for the betterment of humanity, consumed by a magnificent obsession for discovery and caring for little else.

Of course there are individual archaeologists who would fit this bill, or at least satisfy some of these points; but the full picture is rather different. Archaeology today is primarily about careers, and as Campbell notes succinctly, careers are built on results. Personal ambitions override sound research designs, and a complex interplay of negative factors, including a "nasty hierarchy", determines direction. A preliminary epistemological analysis of archaeology, i.e. an examination of how it acquires and interprets its claims of knowledge, suggests several areas of concern. First, its interpretations are generally not testable, hence it cannot be regarded as a scientific pursuit. Second, it is historically prone to mistakes, perhaps more so than any other discipline or academic pursuit (as we will soon see). Third, its paradigm is determined by consensus or majority decision, which is guided very much by prestige and academic weight (the 'silverback phenomenon': assertive alpha males determine dominant models). Fourth, it does not take kindly to being corrected; in fact it treats dissenters badly. And it is particularly repressive, even callous, when the dissent comes from scholars who are not recognised as professional members of the discipline. An example is the *Valetta Convention* in Europe, which seeks to outlaw amateur archaeology on the pretence that it is damaging to archaeological monuments, when in fact dependent archaeologists (those working for the state) may be the principal threat to archaeological resources, in the form of pathological practitioners.

Other dimensions of the discipline are its various ambiguities. For instance, it both supports and opposes the aspirations of indigenous peoples relating to cultural heritage. It creates taxonomies or systems of material evidence, but there is no evidence that these are valid reflections of reality. It makes extensive use of the sciences and seems to have aspirations of becoming a science, yet it maintains a non-scientific epistemology by rejecting principles of falsifiability. Archaeology values its material evidence and jealously guards it, yet it is also the most effective destroyer of this evidence. In fact it destroys nearly all evidence — not intentionally, one might say, but because it lacks the methods and understanding it has yet to gain (e.g. sediments are always destroyed by excavation, and more than 99% of the information available from them is discarded in the process; or by excavating bones and placing them in a collection, the destruction of their DNA is greatly accelerated; Pruvost et al. 2007); and there are countless similar effects, many of which we cannot as yet understand. The most important technique of archaeology is excavation, resulting in the creation of recordings supposedly depicting the stratigraphy of the sediments, and yet there is no facility to test the suppositions made by the recording researcher; the strata no longer exist. In the final analysis, archaeology cannot even be described as a discipline. The only discipline it exercises is consensus, and if we removed from it every area of research that effectively belongs to another discipline or field (geomatics, statistics, sedimentology, nuclear physics or rock art science, to name just a few), archaeology turns out to consist of very little autonomous knowledge; in fact excavation technique is its only major disciplinary asset.

The points raised here are only preliminary, there are more fundamental, epistemologically debilitating factors to consider. They will emerge in due course as we begin to examine the various philosophical or theoretical models that have dominated archaeology, and that have determined the direction of the discipline historically. This is the task of the next chapter.

## REFERENCES


Arcà, A., R. G. Bednarik, A. Fossati, L. Jaffe and M. Simões de Abreu 2001. Damned dams again: the plight of Portuguese rock art. *Rock Art Research* 18: i-iv.

Bahn, P. G. 1990. Motes and beams: a further response to White on the Upper Paleolithic. *Current Anthropology* 32: 71–76.

Bednarik, R. G. 2006. *Australian Apocalypse. The story of Australia's greatest cultural monument.* Occasional AURA Publication 14, Australian Rock Art Research Association, Inc., Melbourne.

Bednarik, R. G. 2007. The science of Dampier rock art — part 1. *Rock Art Research* 24: 209–246.

Bednarik, R. G. 2008. More on rock art removal. *South African Archaeological Bulletin* 63(187): 82–84.

Bednarik, R. G. 2011. Rendering humanities sustainable. *Humanities* 1(1): 64-71; doi:10.3390/h1010064; *http://www.mdpi.com/2076-0787/1/1/64/*

Bednarik, R. G. 2012. The origins of modern human behavior. In R. G. Bednarik (ed.), *The psychology of human behavior.* Nova Science Publishers, New York.

Binford, L. R. 2000–2001. On science bashing: a bashful archaeologist speaks out. *Bulletin of the Deccan College Post-Graduate and Research Institute* 60–61: 329–335.

Bustamente Díaz, P. 2006. Rock art destruction at El Mauro, Chile: one of the world's largest mining waste dams. *Rock Art Research* 23: 261–263.

Campbell, F. 2006. Molesting the past. *The Weekend Australian*, 25 February: R15.

Consens, M. 2006. Between artefacts and egofacts: the power of assigning names. *Rock Art Research* 23: 79–83.

Dyson, S. L. 1997. Archaeology de-damned. *Archaeology* 50(1): 6.

Elia, R. J. 1996. A seductive and troubling work. Review of C. Renfrew, 'The Cycladic spirit: masterpieces from the Nicholas P. Goulandris Collection'. In K. D. Vitelli (ed.), *Archaeological ethics*, pp. 54–61. Altamira Press, London.

Escobar, A. 1991. Anthropology and the development encounter: the making and marketing of development anthropology. *American Ethnologist* 18(4): 658–682.

Frankel, D. 1993. The excavator: creator or destroyer? Antiquity 67: 875–877.

Freeman, L. G. 1994. The many faces of Altamira. *Complutum* 5: 331–342.

Habermas, J. 1979. Communication and the evolution of society, transl. T. McCarthy. Heinemann, London.

Hobsbawm, E. J. 1992. Ethnicity and nationalism in Europe today. *Anthropology Today* 8(1): 3–13.

Houtman, G. 2006. Double or quits. *Anthropology Today* 22(6): 1–3.

Houtman, G. 2007. Response to Laura McNamara. *Anthropology Today* 23(2): 21.

Kohl, P. L. and C. Fawcett 2000. Archaeology in the service of the state: theoretical considerations. In P. L. Kohl and C. Fawcett (eds), *Nationalism, politics, and the practice of archaeology*, pp. 3–18. Cambridge University Press, Cambridge.

Lewis-Williams, J. D. 1993. Southern African archaeology in the 1990s. *South African Archaeological Bulletin* 48: 45–50.

McNamara, L. A. 2007. Culture, critique and credibility. *Anthropology Today* 23(2): 20–21.

Price, D. 2000. Anthropologists as spies. *The Nation* 271(16): 24–27.

Price, D. H. 2005. America the ambivalent: quietly selling anthropology to the CIA. *Anthropology Today* 21(5): 1–2.

Pruvost, M., R. Schwarz, V. Bessa Correia, S. Champlot, S. Braguier, N. Morel, Y. Fernandez-Jalvo, T. Grange and E.-M. Geigl 2007. Freshly excavated fossil bones are best for amplification of ancient DNA. *Proceedings of the National Academy of Sciences of the U.S.A.* 104(3): 739–744.

Silberman, N. A. 2000. Promised lands and chosen peoples: the politics and poetics of archaeological narrative. In P. L. Kohl and C. Fawcett (eds), *Nationalism, politics, and the practice of archaeology*, pp. 249–262. Cambridge University Press, Cambridge.

Simpson, E. 1997. *The spoils of war — World War II and its aftermath: the loss, reappearance, and recovery of cultural property*. Abrams, New York.

Trigger, B. G. 1984. Alternative archaeologies: nationalist, colonialist, imperialist. *Man* 19: 355–370.

Trigger, B. G. 1989. *A history of archaeological thought.* Cambridge University Press, Cambridge.

# 2

# VERSIONS OF ARCHAEOLOGY

In the first chapter we have briefly reflected on the notion that the concept of what archaeology is can differ significantly, for instance in different parts of the world. Here we will examine this proposition in greater detail, and explore the reasons for having so many different archaeologies. After all, the same does not apply to other disciplines: the concepts of chemistry, biology or geology are broadly the same, anywhere on the planet. Regionalisation of disciplines is more apparent in what are called the 'social sciences', and this becomes particularly noticeable in those dealing with history or History (the capitalised form refers to what one privileged section of humanity decrees represents history). Archaeology is perhaps the most fragmented field in this sense.

In principle the archaeologies of the world can be divided by three fundamental criteria: geographical/political differences, theoretical/ideological varieties, and by specific subject preferences. The first criterion refers to the significant differences existing between regional or national schools of thought, the socio-political contexts in which these operate, and the expectations of the respective political masters as well as the public. The second criterion addresses the significant differences between the underlying theories guiding these efforts, as well as the deliberate or subconscious applications of ideological or perhaps subtle political notions. Finally, there is a third criterion, referring to the specialisations that have emerged, such as industrial archaeology or numismatics, to name just two of many. Obviously these criteria can lead to endless combinations or avatars, such as, for instance, a feminist-inspired interpretation of post-processual archaeology focusing on the 'Neolithic Revolution', which may lead to a comprehensive interpretation of a body of (supposedly neutral) 'data' that would be very differently interpreted by, say, a Marxist bias in a postmodern framework. Therefore the possibilities of combining different interpretational criteria seem almost unlimited, and each such combination is likely to result in different models of what a specific set of data mean. This is particularly evident when we consider that the data themselves are not neutral or objective; they were collected according to the predispositions of the researchers concerned, which would be of the same great range. Therefore it is reasonable to expect that there should be as many archaeologies as there are commenting archaeologists.

The most ink has been spilt about the different theoretical or ideological branches of archaeology, so we shall examine them first.

## Theories of archaeology

### *Traditional notions*

Historically, archaeology as a discipline emerged especially in the first half of the 19th century, partly as a development of previous antiquarianism, partly in response to the demand of the emerging nation states after Napoleon to create origins myths justifying their existence. Antiquarianism first appeared in the early 15th century as a concern of the developing humanism of the Renaissance. The first 'archaeological' work is attributed to the 18th century, particularly to William Stukely who used formal surveying techniques to record monuments. By the early 19th century, interest in the past led to systematic quests in various parts of the world, but it also introduced the first fundamental controversies, which has remained the dominant pattern to the present time. We will visit some of these controversies in Chapter 4.

Prior to the 1960s, archaeological theory was dominated by the idea that culture was normative, i.e. that artefacts are expressions of cultural norms, including what Richard Dawkins has later called 'memes' (or Semon's 'mnemic traces', 72 years before Dawkins; see Semon 1904, 1921: 24). Therefore one excavated 'cultural layers', created a taxonomy of their material finds and compared it with the contents of other layers to determine what was regarded as the geographical extent of a culture. Similarly, if such cultural traits were seen to move geographically through time, this was often regarded as evidence for the movement of the carriers of the culture in question, an ethnic group perhaps (e.g. Childe 1929). So for instance in Europe we invented a *Glockenbecher*-culture on the basis of the frequent occurrence of small ceramic beakers 'identifiable' by their bell shape (Figure 1). It dates from about 4500 to 3800 BP, spanning thus the final Neolithic to the early Bronze Age, and an itinerant 'beaker folk' was invented as its carrier, an ethnic group that 'invaded' a variety of regions at various times (e.g. Abercrombie 1902; Harrison 1980). More recent interpretations of the same data favour a social explanation of the phenomenon, involving no mass-movement of people (Lanting and Van Der Waals 1972; Sherratt 1994), but a movement of ideas ('memes'), perhaps about status.

*Figure 1. Bell beakers of the mid-Holocene, Sweden.*

The same kind of colonisation logic based on fetishisation (objects come to represent something else, namely people) is widely found in archaeology, for instance in the concept of a Celtic nation-like people. Today we find it entrenched in the idea that the apparent diffusion of genes over enormous time spans, e.g. from Africa to Europe, proves the mass-movement of an ethnic group in the Late Pleistocene. This shows that archaeology can correct an error, by more careful interpretation, but in time goes on to use the same erroneous logic in a different context.

This traditional approach to interpreting the human past perceived culture as polythetic: its identification requires the co-occurrence of a number of traits, and it was the archaeologist's task to identify and systematise these traits. This leads to particularisation, or an emphasis on differences rather than similarities. Cultures need to be perceived as relatively unchanging chunks of shared ideas and ways of doing things. These can expand or move through the landscape, indicating migrations of people that either supplanted others or colonised areas not occupied by other tribes. This notion of pre-History essentially perceived a timetable filled with 'cultures' indicating both their occurrence and their movements. Accordingly, archaeology consisted of creating the taxonomies that make the identification of cultures and their movements possible.

*The 'New Archaeology'*

The 1960s witnessed the first sustained challenge to this traditional view of the task of archaeology. The principal prompter of this movement, Lewis Binford (1964), described the previous model as "an aquatic view of culture": in it the pre-Historic world was like a pool of water, in which ripples were caused by stones (innovations) dropped in it, leading to interaction of these ripples (Figure 2). The New Archaeology demanded that the discipline must be more scientific and more anthropological. The second demand is related to the fact that in the United States, archaeology is regarded as a sub-discipline of anthropology, while the first refers to the dissatisfaction with tangible progress. Science was seen as progressing with time, whereas traditional 'culture history' seemed

*Figure 2. Professor Lewis R. Binford.*

to be static, simply accumulating more 'data' about fetishes. The reason for science's progress is its practice of testing hypotheses, of seeking falsification. But therein already lay the seeds of the demise of the New Archaeology: that field can import scientific propositions from other disciplines, but it is itself inherently unscientific (not susceptible to falsification). Similarly, the view of culture as "man's extrasomatic means of adaptation" (Binford 1964) is doomed to failure: it presupposes that humans adapt through culture, whereas other animals do so through their bodies. This notion is a fundamental but widespread error in archaeology. Other animals, too, have culture, because the scientific definition of culture is the passing on of practice by non-genetic means, i.e. learning (Handwerker 1989). And a large range of learning occurs in the animal kingdom, as does niche construction and the domestication of other species.

Nevertheless, the more positive aspects of the New Archaeology need to be made explicit. There was the demand to consider one's biases and to avoid simple intuition and implicit assumption. The concept of research design was given much more attention: what were the specific questions to be addressed, how would one test specific hypotheses? Perhaps more importantly, traditional archaeology had tended to focus on the more spectacular aspects of the past, the elites of civilisations, whereas the New Archaeology sought to secure more representative sampling to lead to more systematic description of past societies.

However, there is only limited consistency in the new approaches of the 1960s, which were soon lumped together with what came to be known as processual archaeology. This began formally with Flannery's (1967) argument that 'culture process' was the true aim of archaeological research. The archaeologist ought to search for the systems or mechanisms, be they geological, ecological or social, which brought into existence the patterns in which the archaeological record presents itself. This led to a new emphasis on experimental or replicative archaeology, and to a new field called ethnoarchaeology. 'Analytical archaeology' (Clarke 1968) and 'functionalist' approaches (Binford 1972) became new buzzwords, as did Binford's (1981) 'middle-range theory'. The latter's most outstanding contribution to archaeology has perhaps been the revision of the interpretation of various forms of evidence that had previously been interpreted as cultural, but which were now seen as the result of site formation processes. This was the first introduction of taphonomy into archaeology, a science that had been in use in palaeontology for about forty years at that time (Efremov 1940). Not only is this an indication that underlying principles of other disciplines were adopted very slowly, to this day most archaeologists have an inadequate understanding of taphonomy. Most still think today it is something to do with bones, when in fact it applies to all aspects of archaeology (Bednarik 1994). Moreover, archaeologists tend to view taphonomy as actuo-palaeontology, which ironically is precisely what Efremov sought to replace (Solomon 1994).

It is a common feature of archaeology to reluctantly import ideas from other disciplines, and to then misapply them (see example in Chapter 8).

With middle-range theory New Archaeologists had hoped to build a platform of secure statements about the human past from which to infer and test theories. It has not been the success it promised to be, and by the late 1980s it was considered to have introduced a narrow scientism into the discipline (Shanks and Tilley 1987a, 1987b). As Clarke (1978: 465) had observed perceptively, the use of scientific techniques "no more make archaeology into a science than a wooden leg turns a man into a tree". Archaeology seemed to lack the most basic requirements of a science: there is no obvious way of satisfying the need for predictability and testability, and the independence from value judgments is hard to envisage in a discipline that is invariably humanistic and political. The experiment of creating a scientific archaeology had failed, and processual archaeology was largely replaced after it had reigned for merely two decades.

More specifically, these models failed because a uniformitarian analogy, the basis of middle-range theory, cannot validly test a proposition. There is no plausible reason to assume that all groups of people go through the same phases of cultural evolution, or that similar rules of their development should apply. In fact we can safely assume that this is not the case and that cultural development has no resemblance of biological evolution. In fact the confusion is evident in applying the concept of evolution, an utterly dysteleological process, to the process of cultural change, which may firstly be teleological, and secondly allows for 'devolution'. Similarly, an offshoot called 'behavioural archaeology' seeks to investigate how artefacts become deposited in archaeological sites, i.e. the patterns of use, discard and recovery. Again, taphonomy is the key to interpretation, but the epistemology is similar to that of middle-range theory, presenting the same severe limitations.

Yet another offshoot of processual archaeology is called cognitive processualism. Often confused with cognitive archaeology (a broad and non-prescriptive endeavour of enquiring into the cognitive development of Pleistocene hominins), it seeks to identify behaviour related to past belief systems, to cosmology, religion and ideology. 'Structuralist archaeology' is one more facet of the 'New' Archaeology, which today is a rather 'old' and tired model. Structuralism sees culture as a kind of language, i.e. based on a set of implicit 'grammatical' rules. To understand the system of a culture, one needs to explore the hidden rules that have generated the ways in which culture was externalised. Artefacts, in this system, express structured worldviews, so to examine a binary ideology one would look for consistent oppositions in the presentation of artefacts. For instance the attempts by André Leroi-Gourhan to explain the syntax of Pleistocene cave art in France and Spain by proposing male-female structures are a practical application of these principles. However, because of the extensive contradictions they are hardly accepted today, and it is in any event obvious that such structures cannot be effectively tested.

One more distinctive branch of essentially processualist archaeology is the one informed by Marxism. Its greatest strength is that it links archaeological interpretation to politics, an undeniably valid connection to make. Here, as in structuralism, the question is not so much about the cogency of the theory, but about how to translate its message into testable archaeological propositions. There can be little doubt that modern Western constructs of the world are ideological systems legitimising capitalism, and thus unsuitable as the basis of scientific enquiry. Similarly, there can be no reasonable doubt that archaeology is primarily a political discourse, as we have seen in Chapter 1 and will revisit in Chapter 3.

*Postprocessual archaeology*

This archaeological movement grew out of dissatisfaction with processual archaeology in the early 1980s, and also as a vehicle for a new generation of practitioners to create their own academic niches. It represents a distinctive retreat from the demand that archaeology become more like a science, in fact it explicitly acknowledges that the discipline cannot be a science. Instead the many different factions of postprocessual archaeology claim that all data are inevitably theory-laden. Perhaps that is true of archaeology as it is being conducted, it is not correct of proper sciences: the proposition of the periodic table of elements, for instance, is not theory-laden; attempts to falsify its principles are unlikely to succeed. Another admission inherent in postprocessual archaeology is that archaeologists interpret what they find, and they assume that their hermeneutic interpretations are like those of the people of the past. Therefore one trait shared with Marxist archaeology is the acceptance that the meanings produced by archaeologists are the results of political acts — the imposition of present-day values on the evidence. Postprocessual approaches are often called 'contextual archaeology', or they are referred to as 'interpretative archaeologies' (note the use of the plural). There is also an emphasis on the need to consider the values of the past, or the thoughts of the people being studied, and in that sense a similarity with the aims of cognitive processualism might be evident.

Judging from the results the experiment of postprocessualism seems to be a questionable improvement on the preceding theory. One of the examples sometimes cited refers to the interpretation of rock art. Chris Tilley, a main protagonist of this school, has conducted a much-mentioned 'analysis' of petroglyphs at Nämforsen in Sweden (Tilley 1991). He calls them 'rock carvings' (made with a carving knife, perhaps? Swedish petroglyphs are usually the result of percussion) and begins by telling the reader that they are of the third millennium BCE — when in fact no rock art in Scandinavia has been satisfactorily dated. He constructs an interpretation of an art of which he only has mediocre interpretations (much of the rock art had been destroyed since it was recorded almost a century earlier), only to then deconstruct it himself, and telling the reader that it is impossible to determine the meaning of the art. So the question arises, why write a book that merely interprets someone else's interpretations, and then simply tells us what we have already known since Macintosh's (1977) landmark study: that emic meaning is not recoverable in rock art, except with the help of experts (indigenous people possessing emic access to the corpus in question).

A side branch of postprocessual archaeology is the archaeology of gender. This is an openly political form of archaeology, concerned essentially with two principal issues: an opposition to androcentric terminology, ideology and interpretation; and an opposition to sexism in recruitment, funding and promotion within the academic establishment (i.e. it is a self-promotion of archaeologists). The far more interesting aspect of feminist archaeology concerns the proposition that 'rationality' is an androcentric conflation that biases science in favour of male ways of seeing the world.

*Postmodernism in archaeology*

Essentially, postmodernism as applied in referring to archaeological theory seems to be a return to relativism. There is a link with the neo-pragmatism of American philosopher Richard Rorty, and an interest in the rhetorical regress or self reflexivity of allowing language to become the object of its own scrutiny. In the applications of this vague theory to archaeology we seem to have come full circle. Indeed, the idea of relativism has been around at least since Protagoras ('Man is the measure of all things', which is both valid and absurd — as already shown by Socrates). Nietzsche and Wittgenstein argued against reason, as did Feyerabend (Figure 3) more recently, and postmodernism is really a well-established epistemological stance made once again fashionable in the 1990s.

*Figure 3. Professor Paul Feyerabend.*

In contemporary archaeological theory, postmodernism has produced conflicting messages. On the one hand, disciplinary boundaries need to be broken down; on the other there is a fragmentation of method and multitude of new approaches that lead to more niche archaeologies. By the same token, postmodernism decrees there can be no neutral method, which means that alternative views also have validity, reflecting Feyerabend's 'Anything goes!' dictum. But if that were the case, New Age archaeologies and 'folk archaeologies' would have as much justification as any other version of the past. If we accept this position we also accept that university-run archaeology has no privileged academic position, it is simply one of competing stories of human history, and there is no *objective* way of assessing any of them.

It would appear that postmodernism was introduced at an inopportune time for archaeology. Weakened by its lack of underlying universal theory, by its endless internal divisions and epistemological contradictions, by its inability to secure scientific standing, by its wrangling with the indigenous owners of cultural heritage and by the many other factors we have visited in Chapter 1, it now faces an intellectual climate that is sceptical, even hostile, to finite claims of knowledge. It is truly a discipline in crisis, unprepared for such an assault on its authority. Archaeology has little or no comprehension of its own theoretical underpinnings, debate of archaeological theory is "of a very low intellectual standard" (Johnson 1999: 182); endeavours to be scientific are defined as "vacuous" and the attempts of their presenters at learning as "pathetic" (Binford 2000–2001: 334). The new relativism of postmodernism haunts all of the 'social sciences' and their lack of consilience with the real sciences, but if it were allowed further inroads into archaeological thinking it would logically lead to the dissolution of the discipline. In the final analysis, archaeology has a choice between science and postmodernism. The next decades will show how it responds. In all probability it will turn back towards science, and the cycle will begin anew: archaeology is like a dog chasing its own tail.

## Regionalism in archaeology

One of the many fallacies of archaeology is the implicit assumption that there is some kind of 'world archaeology': a universal body of knowledge, data and methods that is shared by all archaeologists of the world. In reality, the underlying notions of the purposes and extent of the discipline, its methodology, the available knowledge base, the relevant political imperatives and many other factors differ substantially between geographical regions. Although the term 'archaeology' is shared worldwide, some practitioners see themselves as prehistorians, whereas a great deal of archaeology is not even concerned with 'prehistory'. Indeed, as noted in Chapter 1, that term is itself inappropriate, because it refers to a time before what one privileged academic enclave of humanity considers being 'history'. The term would be offensive to most humans who have ever lived on this planet, and who were not able to read or write. That applies to some people right up to the present, and all of them experienced their times as history, not as prehistory. To separate 'history' from 'prehistory' on the basis of such a variable as writing is absurd; there is no scientific evidence that written history is more reliable than oral, nor is that proposition testable. In fact it is easy to construct an argument that the opposite is true, that oral transmission is more reliable than written (cases in Australia pertain). Besides, most people in history could not write (many still cannot today) and one would need to decide whether such people as the Mayas were 'historic' or 'prehistoric' (cf. Bouissac 1997). To qualify the use of 'history' by referring to a specific period of time marked by the advent of a specific feature, it is best to capitalise the word as 'History', creating a named entity, and derive from this the term 'pre-History'. It is obvious that there cannot be a history before history; hence there can be no prehistory.

The confusion is best illuminated by remembering that in German, the word *Geschichte* has two entirely different meanings (cf. *storia* in Italian). In the first, it translates as 'history', in the second as 'story', which is most appropriate: history is in reality a story, a narrative about the past, an interpretation punctuated by factual snippets, and always heavily edited by many processes (including the fact that for most of written history, most of the people living then were illiterate, so history was inevitably written by the winners, by elites and their scribes; some Andean rulers took this principle so far that they executed all historians after they acquired power). Therefore we have no credible evidence that 'prehistory' is less reliable than history, because we are comparing two sets of stories that have both unknown levels of veracity.

To consider the regional differences of archaeologies we could begin by reviewing those between the United States and Europe. There are very few archaeology departments in American universities, most archaeology there is attached to anthropology departments. In a general sense, archaeology refers primarily to the past of 'other' cultures, whereas in Europe pre-History is seen as the extension of History, the recent historical period we tend to identify so because it has left us written records. More than half the archaeologists specialise in the periods of History, and more than half the excavations take place at Classical or post-Roman sites. Yet in Europe, iconic pre-Historic sites such as Lascaux or Stonehenge are seen as quintessentially French or English monuments, when in fact they are clearly not the work of identifiable French or English people. Americans, by contrast, view their archaeological monuments as someone else's cultural property. The same applies in Australia, but in China or India, for instance, the European attitude pertains. As a general rule, the nations politically dominated by recent colonising groups tend to subscribe to the American view. In the case of Australia this has led to the contradictory situation that archaeologists copy English archaeology to the letter, but then eschew the English attitude to archaeological heritage by regarding most Australian archaeological heritage as 'someone else's'.

To some extent these differences are historically determined, but political ideologies and priorities often influence the structure and direction of regional archaeologies. For instance in China, early human history is regarded as palaeontological rather than archaeological, while the rock art of the 'historical' periods tend to fall under the aegis of art history rather than archaeology. The former Soviet Union, another socialist system, also applied the principles of Marxism in preference to capitalism, the ideology of the West. These political frameworks had significant effects on the way archaeology was conducted, and despite the recent weakening of the socialist world, profound differences remain between regional archaeologies that are the outcome of political ideologies. As we have seen in Chapter 1, these range from regions dominated by distinctly nationalistic regimes to those exercising apartheid or caste systems or those subjected to dictatorships. To suggest that their respective systems of archaeology are compatible, or closely relate to some fictitious notion of a universal world archaeology would be naive.

Similarly, archaeology in most parts of the world — socialist countries being the obvious exception — is strongly influenced by religions. To illustrate with an example, until quite recently, rock art was severely neglected in most Islamic countries, simply because the practice of depiction was frowned upon as being blasphemous. This position has been abandoned only in recent years, for instance by the new practices introduced in Saudi Arabia, which are right now influencing those of countries such as Iran and Pakistan. In Saudi Arabia we had the absurd situation that the only four major works about the country's very extensive rock art were all written by one researcher, Emmanuel Anati, an Italian who had never in his life been to that country. This has recently been corrected through the work of Majeed Khan and others, and I have had the opportunity of being involved in these reforms (Bednarik and Khan 2005). Today Saudi Arabia has one of the world's best practices of rock art site management and protection, and the rest of the Islamic world is expected to follow this example.

To appreciate the degree of influence religions exercise on archaeological practice, one could consider many other examples, some being subtler, others perfectly straightforward. It is obvious that a great deal of archaeological work is funded and conducted by religions, directly or indirectly; for instance in the United States or in Israel. Its purpose is clearly not an idealistic and innocent search for what really happened in the past, but the systematic reinforcement of pre-existent belief systems through securing 'evidence' that confirms these. Again I can cite an example from my own experience.

*Figure 4. Fake inscription on limestone sarcophagus purporting it to be of Jesus's brother.*

Some years ago I was asked to consider coming to Israel to attempt scientific dating of an inscription on a stone coffin purported to be the sarcophagus of Joseph, brother of Jesus (Figure 4). I found a convenient excuse to keep out of what I regarded as a ploy by Christian fundamentalist archaeologists. Soon later I learnt that Israeli archaeologist Yuval Goren had conducted excellent detective work on this case, and had traced the fake coffin to a major operation — a whole factory producing archaeological forgeries on an industrial scale. There are thriving industries in many parts of the world churning out millions of fake archaeological objects, and the Middle East is perhaps the foremost supplier for this world market. Whereas in the rest of the world, profit is the only motive of these industries, in the Middle East there are religious overtones, nourished by religious fundamentalism. All major religions have a distinctive interest in archaeology, extending back to the very beginnings of the discipline, but the same can be said about New Age movements of all kinds of persuasion. In recent years there has also been a renewed interest in archaeology by fundamentalists trying to disprove the idea of human evolution, which is reflected in countless debates on the Internet.

Mainstream archaeologists oppose all of these manifestations of 'folk archaeology', but not with quite as much enthusiasm as one might expect. While these fringe interests do run counter to preferred paradigms, they also serve to pique the interest of the public in the products of archaeology, and they also help to reinforce and justify the role of archaeologists as 'experts' or adjudicators of matters of the human past.

All of these factors so far listed contribute to the fragmentation of archaeology and to its heterogeneity, and to the development of distinctive traditions. But even more influential in the creation of regional differences may be the inherent preoccupations of archaeologists. These are not only determined by the many factors already mentioned, but most especially by local aspects of special archaeological interest. For instance in Egypt there is such focus on the Pharaonic periods that all other subjects are neglected, and similar preoccupations can be found in many parts of the world, including Mexico, India and the Middle East. The effects of this can be so great that such local schools of archaeology bear hardly any resemblance to the concept of a 'world archaeology'. For instance in the Andean countries, specialisation in the relatively recent Andean civilisations can be at the expense of any other archaeological knowledge. I recall working with a senior professor of archaeology in Peru who noticed my interest in specific stones in the field. His knowledge of the ceramic periods of Peru was encyclopaedic, but he had never encountered the idea of stone tools, and was amazed by my interest in the countless hand-axes I saw on Pleistocene river terraces. It had never occurred to him that these were tools. To me such artefacts were totally out of place in this region, and I realised that the reason why they had remained unreported was that local archaeology had not developed a sustained interest in the pre-ceramic periods.

On another occasion I was shown the famous rock painting sites of Raisen in central India, accompanied by a group of local archaeologists, who had all been to the sites before. As we entered the first shelter, everyone looked up at the brilliant paintings, and camera shutters began to click. Then I drew attention to a group of petroglyphs on the floor of the shelter and my friends were amazed: they had either never noticed them, or had not realised what they were. They had seen the word 'petroglyph' in print, they had no doubt seen many petroglyphs — but without making the connection between the name and the phenomenon. This occurred in 1990 and it was the moment of the discovery of petroglyphs in central India. Today there are hundreds of petroglyph sites known in Madhya Pradesh and Rajasthan, and they include the two oldest known rock art sites in the world.

These anecdotes show that to assume the existence of a universal body of knowledge, data and methods that is shared by all archaeologists of the world is entirely unwarranted. One powerful separating factor is language. Every time key knowledge long available in other languages penetrates into Anglophone mainstream archaeology, which is largely monolingual, there is excitement as if a new discovery had been made. For instance the Schöningen Lower Palaeolithic spears from Germany had been long known, but had only been published in German. When they were first reported in Britain the announcement was treated as a breakthrough; yet it was so only for those on the English-language side of the language barrier (there are in fact seven European sites with reported wooden spears of the Lower Palaeolithic). Numerous examples of this phenomenon could be cited, and it adds to isolation and fragmentation in the discipline.

Archaeology is, as an eminent South African professor of archaeology stated, what local senior archaeology professors say it is (Lewis-Williams 1993). And the differences between many of these regional archaeologies could not be greater. The notion of a uniform global concept of archaeology is a myth, and effective communication between many of these schools is almost impossible. They may ritually meet at international conferences and they do communicate, but still mostly within specific enclaves. There is very little constructive dialogue with other dioceses of archaeology, or between their respective chief shamans. This is not only about significant differences in knowledge, data and method, it is also very much about territory and what goes with it: prestige, influence and funding. What all these hundreds of specialisations have in common is a vague commitment to the study of past cultures or peoples, and a notion that this is achieved primarily, but certainly not exclusively, through the method of excavation. Clearly the common denominators are not adequate to treat this as a uniform discipline. Rather it comprises a large number of constructs based on a loosely shared curiosity about the human past — a collection of vocational archaeologies of greatly varying knowledge bases, ideologies and methods.

## Subject specialisation in archaeology

At this stage it has already become apparent that universal world archaeology remains an elusive ideal, a fantasy, and yet there is one more major factor dividing the discipline into numerous factions. Throughout the course of the 20th century, archaeology and 'prehistory' became progressively fragmented into ever-increasing numbers of sub-disciplines and specialties. The process began in the late 19th century, especially once the notions of both a Palaeolithic period and a Palaeolithic rock art had become established in Europe, as well as the division into basic periods (the Stone, Bronze and Iron Ages, introduced by Christian Thomsen, first publicised in 1819) had been generally accepted. By the end of the 20th century, the good number of métier archaeologies had mushroomed. The "octopus of archaeology" (Lorblanchet 1992) included a vast range of 'specialisations', ranging from Pliocene hominin evolution to 'historical archaeology', from numismatics to 'garbology'. The separation of these subject-derived boutique archaeologies from the above-listed divisions according to their underlying theories had become increasingly blurred. Many of these specialisations made no use of excavation as a method, and many, perhaps even most of them, no longer dealt with pre-History, but focused purely on recent periods, including the present. As these niches expanded their spheres of influence, their priorities often developed away from the initial scope of archaeology. For instance nautical or maritime archaeology is now almost exclusively concerned with sunken shipwrecks, so its activities revolve around diving more than around the understanding of the distant human past. The initial exploration of the oceans in the Pleistocene, one of the most exciting chapters of human history, has attracted no interest whatsoever from maritime archaeologists, and all publications addressing this subject, in any language, have been written by just one researcher, myself. Maritime 'archaeology' is now entirely focused on the mapping and recovery of wreckage, most of it of recent centuries. Its results can tell us very little about the development of ancient people, or influence our ideas of human evolution. In a scientific sense, these efforts can only yield trivial results: details about some ship of which we have in any case historical records, perhaps about why or where it sank. The underlying impetus for this rather expensive pursuit comes from the motives of treasure hunting (and many of these wrecks do contain treasure), so the sub-discipline derives its appeal primarily from an image of adventure and news-making. Since it focuses mostly on the very recent past, as do numerous other specialty archaeologies, it can no longer be said to be concerned with ancient cultures (contra Fagan 1998: 4), and its inclusion with archaeology rather than history is perhaps based on its occasional use of a modified form of excavation. But this is an inconsistent use of the term archaeology: excavation is not a method used by all branches, nor is it exclusive to archaeology; therefore it does not define an archaeological sub-discipline. It is not, for instance, involved in rock art research or aerial archaeology. Nor is there any reason why excavation is not an admissible method of history. Some, such as Fagan, would argue that those branches that deal with recent or historical periods would be better accommodated in the discipline of History, but then the question arises, where does one set the chronological separation. Most practitioners see Greek or Roman history as legitimate concerns of traditional archaeology.

Another example of these contradictions is provided by the Tucson Garbage Project Bill Rathje has initiated. It can be seen as a form of 'behavioural archaeology', in the sense that it investigates how and in what forms present people in Arizona discard rubbish. It is contended that the patterns of contemporary garbage disposal might inform us about what happened in the patterns of discard of past peoples. This project would employ the method of excavation and sampling as typically used in archaeology, and its distinctively middle-range purpose is to illuminate human behaviour in the distant past. Industrial archaeology, modern settlement archaeology or any other form of historical archaeology, on the other hand, are clearly not concerned with the interpretation of the distant past by analogy, and their inclusion in the discipline seems inappropriate: they should be sub-disciplines of History.

If we combine the effects of the fragmentation by theoretical underpinnings, countless regionalisms and different subject specialisations we arrive at an almost endless range of possibilities to divide archaeology into. We might consider what a structuralist feminist mode of Upper Palaeolithic archaeology in western Europe might decide the figurines of that period were used for (Dobres 1992). We might find that alternative interpretations arrive at significantly different deductions (e.g. Duhard 1989; Bednarik 1996; Russell 2006). We might explore the ideology or method of what a processualist Stalinist archaeologist of the former Soviet Union might have revealed about shipwrecks of the American Civil War; or consider the findings of a Brazilian ethnoarchaeologist commenting on the meanings of South African San paintings and their relevance to Hawaiian petroglyphs. If this cacophony of archaeologies is not sufficient to convince us that there is little rhyme or reason in the discipline, we could even introduce the very valid concepts of Trigger (e.g. 1984, 1985, 1989), Silberman (e.g. 1982, 1989, 1995) and others that archaeology is primarily a political pursuit, that it consists essentially of three forms of discourse (nationalist, colonialist and imperialist, see Chapter 1), supplemented by such approaches as those of 'touristic archaeology' and an 'archaeology of protest'. That would really liven up the party, and expose the farcical claims that archaeology is a uniform discipline. Clearly it exercises no discipline, it follows a scientific philosophy only in the sense of Feyerabend's "anything goes". In archaeology, almost any narrative goes, and in that sense it is no different from 'folk archaeology' or the models of the past held by the ethnosciences of traditional indigenous peoples, or by religious fundamentalists. Archaeology, in the final analysis, is, in the way it is

*Figure 5. Early drawing by Sergei Eisenstein, showing the community at Riga as individual and anthropomorphised animals.*

being conducted, an academic free-for-all whose principal purpose seems to be the individual advance of ambitious practitioners (Figure 5). In the words of semiotics professor Göran Sonnesson (n.d.): "If there is some particular archaeological domain of study, it is hard to discern it; and if there is a peculiar archaeological point of view, it is not easy to define". Most of the practitioners are certainly not concerned with 'prehistory', but with 'history' (Andrén 1997: 12). It is not about a specific method, but about methods borrowed from forensics, palaeontology, sedimentology, palynology, anthropology, ethnology and numerous other fields. Not surprisingly it is always a social and not an epistemological construct. It is in the final analysis a glorified hobby that derives its major justification from heritage protection laws — and most of its income from 'creative re-interpretation' of these same laws to suit its principal benefactors, such as developers and large resource companies that wish to destroy such heritage values.

There is adequate space debris already on some planets and on the Moon, and floating around in space generally, to begin an 'archaeology of hominins in space'. It is only a matter of time before the 'discipline' will avail itself of this 'cultural resource' to create yet another 'sub-discipline' demanding research funding for what would be essentially a fairly straightforward but rather pointless hobby.

## REFERENCES


Abercrombie, J. 1902. The oldest Bronze-Age ceramic type in Britain. *Journal of the Royal Anthropological Institute* 32: 373–397.

Andrén, A. 1997. *Mellan ting och text.* Brutus Östlings, Bokförlag Symposion, Stockholm/Stehag.

Bednarik, R. G. 1994. A taphonomy of palaeoart. *Antiquity* 68(258): 68–74.

Bednarik, R. G. 1996. Palaeolithic love goddesses of feminism. *Anthropos* 91(1): 183–190.

Binford, L. R. 1964. A consideration of archaeological research design. *American Antiquity* 29: 425–441.

Binford, L. R. 1972. *An archaeological perspective.* Seminar Press, New York.

Binford, L. R. 1981. *Bones: ancient men and modern myths.* Academic Press, New York.

Binford, L. R. 2000-2001. On science bashing: a bashful archaeologist speaks out. *Bulletin of the Deccan College Post-Graduate and Research Institute* 60–61: 329–335.

Bouissac, P. 1997. New epistemological perspectives for the archaeology of writing. In R. Blench and M. Spriggs (eds), *Archaeology and language I*, pp. 53–62. Routledge, London and New York.

Childe, V. G. 1929. *The Danube in prehistory.* Oxford University Press, Oxford.

Clarke, D. L. 1978. *Analytical archaeology.* Methuen, London.

Dobres, M.-A. 1992. Re-considering Venus figurines: a feminist-inspired re-analysis. In A. S. Goldsmith, S. Garvie, D. Selin and J. Smith (eds), *Ancient images, ancient thought: the archaeology of ideology*, pp. 245–262. Archaeological Association of the University of Calgary, Calgary.

Duhard, J.-P. 1989. La gestuelle du membre supérieur dens les figurations féminines sculptées paléolithiques. *Rock Art Research* 6: 105–117.

Efremov, I. A. 1940. Taphonomy: a new branch of paleontology. *Pan-American Geologist* 74: 81–93.

Fagan, B. M. 1998. *Peoples of the Earth. An introduction to world prehistory* (9th edn). Longman, New York.

Flannery, K. V. 1967. Culture history v. culture process: a debate in American archaeology. *Scientific American* 217: 119–122.

Handwerker, W. P. 1989. The origins and evolution of culture. *American Anthropologist* 91: 313-326.

Harrison, R. J. 1980. The beaker folk: Copper Age archaeology in western Europe. London.

Johnson, M. 1999. *Archaeological theory: an introduction.* Blackwell Publishing, Malden and Oxford.

Lanting, J. N. and J. D. Van Der Waals 1972. British beakers as seen from the continent. *Helenium* 12: 20–46.

Lewis-Williams, J. D. 1993. Southern African archaeology in the 1990s. *South African Archaeological Bulletin* 48: 45–50.

Lorblanchet, M. 1992. Introduction. In M. Lorblanchet (ed.), *Rock art in the Old World: Symposium A of the First AURA Congress, Darwin*, pp. xv–xxxii. Indira Gandhi National Centre for the Arts, New Delhi.

Macintosh, N. W. G. 1977. Beswick Cave two decades later: a reappraisal. In P. J. Ucko (ed.), *Form in indigenous art*, pp. 191–197. Australian Institute of Aboriginal Studies, Canberra.

Russell, P. 2006. Learning from curves: the female figure in Palaeolithic Europe. *Rock Art Research* 23: 41–50.

Semon, R. 1904. *Die Mneme*. W. Engelmann, Leipzig.

Semon, R. 1921. *The mneme*. George Allen & Unwin, London.

Shanks, M. and C. Tilley 1987a. *Re-constructing archaeology: theory and practice*. Routledge, London.

Shanks, M. and C. Tilley 1987b. *Social theory and archaeology*. Polity Press, Oxford.

Sherratt, A. 1994. The emergence of elites: Earlier Bronze Age Europe. In B. Cunliffe (ed.), *The Oxford illustrated prehistory of Europe*, pp. 244–277. Oxford University Press, Oxford.

Silberman, N. A. 1982. *Digging for God and country*. Knopf, New York.

Silberman, N. A. 1989. *Between past and present: archaeology, ideology and nationalism in the modern Middle East*. Henry Holt, New York.

Silberman, N. A. 1995. Promised lands and chosen peoples: the politics and poetics of archaeological narrative. In P. L. Kohl and C. Fawcett (eds), *Nationalism, politics, and the practice of archaeology*, pp. 249–262. Cambridge University Press, Cambridge.

Solomon, S. 1990. What is this thing taphonomy? In S. Solomon, I. Davidson and D. Watson (eds), *Problem solving in taphonomy: archaeological and palaeontological studies from Europe, Africa and Oceania*, pp. 25–33. Tempus 2, University of Queensland, St Lucia.

Sonesson, G. n.d. The quadrature of the hermeneutic circle (Lecture 1, Current issues in pictorial semiotics). *Semiotics Institute Online, http://projects.chass.utoronto.ca/semiotics/cyber/cyber.html*, accessed 17 October 2010.

Tilley, C. 1991. *Material culture and text: the art of ambiguity*. Routledge, London.

Trigger, B. G. 1984. Alternative archaeologies: nationalist, colonialist, imperialist. *Man* 19: 355-370.

Trigger, B. G. 1985. The past as power: anthropology and the North American Indian. In I. McBryde (ed.), *Who owns the past?*, pp. 11–40. Oxford University Press, Oxford.

Trigger, B. G. 1989. *A history of archaeological thought*. Cambridge University Press, Cambridge.

# 3

# NEOCOLONIALISM IN ARCHAEOLOGY

Mainstream archaeology is, as we have seen so far, a discipline bedevilled by many problems. Some are fairly obvious and have now been discussed, such as the curatorial aspirations of the discipline that are perhaps rooted in its antiquarian origins. They have led to numerous confrontations with various interest groups, particularly indigenous populations who object to many practices of archaeology. Moreover, they are also a fundamental deontological stumbling block of the discipline in attaining a scientific status. Exclusive executive control over the resource being studied exists in no other discipline (except for ethical reasons in certain areas of medical research), and since such control is incompatible with the independence and the absence of a vested interest that is the first precondition in unbiased scientific enquiry, archaeological claims for scientific status are paradoxical: it is on the basis of scientific status that archaeology makes its claims of control, and yet it is this control that precludes such a scientific status.

The principal epistemological impediment in archaeology, the difficulties of providing hypotheses with adequate opportunities of refutability, has also been addressed already. Here I will rehearse some aspects that have received less attention, and that relate to an underlying neocolonialist ideology. Eurocentric 'science' postulates that the European way of experiencing reality is the only valid one, and all claims of knowledge, to be scientifically acceptable, must be presented in a form that relates them to this model. There is no allowance for the possibility that alternative systems of scholarship, which have been silenced in past centuries by Western military superiority, might provide valid alternatives. This is odd, considering the discovery in the course of the last century that the fundamental laws of Western science are both logically and experimentally inconsistent. One would think that such ideas as Heisenberg's uncertainty principle would provide Western scholars with enough doubts to cease converting humanity to their belief system.

Let us consider a specific example. In empiricism and in archaeology, time is a non-spatial continuum in which events occur in an irreversible, linear succession, incapable of acceleration or deceleration. Many non-European metaphysical systems have quite different perceptions of time, and today we know that time is not the entity in a non-accelerated frame of reference empiricism expects it to be. Yet nowhere is there any acknowledgement in archaeology or anthropology of the possibility that some indigenous concepts of time might be more realistic than that of empiricism, and whenever archaeologists are confronted by indigenous objections to their naive empiricist interpretations of the past, they defend them by reminding us of their status as the shamans of modern times. Like shamans, they are highly elitist, undergo life-long training and professional rigour, believe in their own power of interpretation and in the potency of the means they use in their quest for truth, and have a great influence over the society they serve. Like shamans, they use severely limited knowledge and understanding of natural phenomena and processes in explaining reality, they produce metaphysical models, derive their power and social status from their activities, and like shamans they are the centre of a subjective belief system.

While archaeologists squabble over whether Aborigines arrived in Australia 40,000, 60,000, 140,000, 180,000 or over 300,000 years ago, and change their consensus every so often, Aboriginal people have all along maintained that they were here since time began. Their metaphysical beliefs, which are so much more plausible and realistic then those of the religious belief system the recent colonisers tried to convert them to, explain vividly their descent from animals, and our DNA is indeed almost immortal. European scientists discovered only in the 19th century that their religious model of human descent was incorrect, that humans are descended from other animals. They still consider this to be an important discovery, which only underlines their complete Eurocentrism. After all, Aboriginal savants had always known this. Aboriginal ethnoscientists had long deduced that the Sun goes clear around the Earth, that the tides are caused by the Moon (Warner 1937; Wells 1964; Hulley 1996), and what causes both lunar and solar eclipses is correctly defined in myths about the Sun-woman and Moon-man (Warner 1937; Bates 1944; Johnson 1998). Aboriginal concepts that there was a time in the distant past when rocks were soft, expressed in the ancient stories of many tribes and probably deduced from volcanic flows and fossil casts, would have been complete heresies in darkest Europe until a few centuries ago. Other scholars cannot understand Aboriginal concepts of how events and phenomena are arranged in time and space, and in the typical fashion of Western righteousness they relegate possibly valid aspects of a cosmography to the status of pure mythology. Recent work by psychologist Lera Boroditsky and linguist Alice Gaby suggests that some Aboriginal tribes such as the Pormpuraawan experience time as moving from east to west, and think about the day after tomorrow as two days to the west. This, like the conceptions of time of many other traditional societies (e.g. the Zuni in North America), differs dramatically from the way time is experienced in Western ontologies. However, rejecting such alternatives is intellectual colonialism (Western science has not explained time adequately), and a guild of self-appointed scientific shamans has no mandate to judge or disparage an alternative, and in some aspects probably superior, interpretation of the past.

## Political dimensions of archaeology

This is an example of overt intellectual colonialism, but there are also more subtle forms in scholarship. In Chapters 1 and 2, we have briefly considered some factors influencing archaeology socio-politically. Historically, the discipline arose from the need to provide developing nation-states with grand 'histories' and origins myths (just as sections of it endeavour to justify religious beliefs by searching for supporting evidence). This dependency on the states (or on religion) remains one of the most characteristic features of archaeology, with states providing nearly all employment opportunities for practitioners. They are either servants of the state, at museums, government agencies or universities; or they work as contractors for corporate or government clients, in which case their existence is entirely dependent on relevant heritage protection legislation.

Nationalist imperatives have remained an integral driving force in the evolution of national traditions throughout the world. Leaving aside the more overt nationalist archaeologies, such as those of totalitarian regimes, less extreme forms can be detected in any of the many regional archaeologies, which inevitably developed as national schools (Silberman 1989; Howsbawm and Ranger 1983). This marks again a fundamental difference between archaeology and the sciences, and is manifestly the result of imperatives of nationalism. A good example is provided by the involvement of archaeologists in creating modern mythologies about militarism, e.g. by researching places of 'heroism'. This is not so much driven by demands of the public, but is in direct response to government policies. For instance in Australia, a recent conservative government developed a distinctive policy of glorifying the country's military past, promoting sites of conflict in other parts of the world as national heritage sites (Kokoda Track in Papua New Guinea, Gallipolli in Turkey), while at the same time rejecting any notion that the only true defenders of Australian soil deserved the same consideration. In Australia, tens of thousands of warriors have bravely faced the rifles and cannons of the British invaders with spears and boomerangs, and fought long guerrilla campaigns defending their territory (Figure 6). Yet their sacrifices remain ignored, in fact they were swept under the carpet and labelled by government leaders a part of a 'black armband history'. Instead of challenging this re-writing of history by the state, as would be the role of academe, the country's archaeologists smartly adapted to the political climate and found ways to facilitate the glorification of Australian militarism: they offered their skills to study battle fields, war graves or anything that would attract government funding.

Thus the symbiosis of militarism and nationalism finds much expression in the archaeologists' eagerness to curry favour with the state. Other political dimensions of the discipline are less tangible or subtler, but just as effective. Two types are of particular interest here. As the colonising European powers

*Figure 6. Aboriginal massacre in Western Australia, in 1868.*

expanded their control over indigenous societies around the world, they were able to force their own ideologies on them. Whenever Europeans made attempts to understand the intellectual basis of the metaphysical constructs of people they considered primitive or inferior, the intent tended to be better control of these subjects. Any accommodation to be made in the relationships always had to be made by the colonised; they were expected to adopt the constructs (Wittgenstein's *Begriffswelt*) of their conquistadors, or be relegated to the scrap heap of history.

More insidious is the academic appropriation not only of indigenous histories, but also of beliefs and metaphysical or social constructs, particularly through the practices of anthropology and ethnography. These disciplines, closely related to, and to some degree interwoven with, the hegemonic project of archaeology, profess to serve the betterment of human knowledge, but they are easily corrupted to serve the ambitions of power elites of many descriptions. For instance anthropology operates today in the service of many covert agencies and unethical organisations (see Chapter 1), in much the same way as some scientists serve unethical tobacco companies or corporate entities patenting indigenous knowledge or copyrighting genetic data. These disciplines work largely for the nation-states, therefore their acquisition of knowledge about the ontologies of indigenous societies can easily become integral to dispossession, diminishment of indigenous values, and gaining power through 'interpretation'. Academe is a power hierarchy and a reward system, and any academic consumption of fragile cultural information occurs inevitably in corrupted form, and in a form that serves the dominant society.

It is simply not possible to translate indigenous metaphysics into forms decipherable within a Western construct of reality (Berger and Luckmann 1966; Pinker 2002) without significantly corrupting it. Any such construct is tied to the language providing its framework (Sapir 1929: 209). But even if that impediment did not exist, the exposure of precarious and very vulnerable belief systems to the glare of academic attention is itself injurious to the societies concerned. Many of their beliefs are restricted, sacred or secret, therefore reviewing them, discussing them under the guise of academic freedom already debases them. And most importantly, it thus becomes a political act. As sometimes happens in biology, an organism may change simply by being examined (e.g. exposed to light), and in a sense this always occurs when societies other than one's own are studied. As archaeology usurps non-Western histories, it contributes to devaluing the societies in question. And this occurs even before we remember that most of its interpretations are, in any case, probably false, as we shall see later.

The topic of pathological archaeology has already been alluded to in Chapter 1, and the destruction by archaeologists of cultural sites, sacred sites or places of veneration well expresses the political power of the discipline. Having acquired some basic knowledge about such places, the archaeologist, as servant of the state, proceeds to facilitate their destruction. Archaeologists are specifically licensed by the state to assist developers to destroy such places; they are the states' gatekeepers of sacred sites of indigenous peoples. The traditional owners of such sites always object bitterly to this destruction (and the imposition by the state of an alien metaphysical system over their traditional), anywhere in the world where it occurs. Whereas in Europe, significant cultural heritage sites are very effectively protected, their destruction by European or Western society in the New World occurs on a daily basis. There can be no doubt that this illustrates the colonialist agenda of the modern states of the Americas and Australia: indigenous heritage is not valued in the way it is in Europe. And therein lies the problem, because at the same time it is argued that universal values should apply. There can be no level playing field for indigenous cultural heritage if it is dominated by European values (Leone and Potter 1992).

To illustrate this with a tangible expression: in the state of Tasmania, the maximum fine for damaging indigenous cultural heritage is $A500; the corresponding fine for damaging European (essentially British) heritage is $500,000. In 2006 I pointed out to the state government that this practice is racially discriminatory. Since then there have been numerous enquiries and submissions into the issue, but at the time of writing this is finally being resolved by introducing new legislation. It is important to note that this anomaly is reflected in the attitude of the general population to cultural monuments. There is an endemic culture of vandalism of Tasmanian rock art sites, one quarter of which have been severely damaged as a result of this glaring disparity (Sims 2006; Bednarik 2006a).

*Figure 7. Massive destruction of sacred rock art sites on an industrial scale, perpetrated by archaeologists, Dampier Archipelago, Australia, 2007.*

In various parts of the world, archaeologists are engaged in the destruction of rock art sites by clearing the sites of the 'art'. This practice robs the rock art of its site, and the site of its rock art (Figure 7). It is an act of ultimate cultural usurpation, of colonisation as a continuation of the process begun when the land was originally appropriated and its resident tribes massacred or 'pacified'. It also amounts to a practice of selling humanity's cultural heritage to the highest bidder, usually multinational corporate players of enormous power. It is therefore equivalent to cultural theft, and where it is facilitated by archaeologists, it is definable as *pathological archaeology* (Bednarik 2006b, 2008; Chaloner 2004; Escobar 1991; Houtman 2006, 2007; McNamara 2007; Moore 1999; Price 2000, 2005; Ritter 2003).

Against all of this stand several arguments: that a 'modern state' has little choice, that it must maintain or improve living standards and therefore the economic exploitation of a region's resources has precedence over the cultural values that are destroyed in the process. Also, it would be argued, the state governs for all, including the indigenous population. This may sound reasonable, although it could be said that often the state fails in governing for the benefit of the indigenes (for instance when it removes their children, as occurred widely in Canada and Australia until quite recent times). Moreover, it must be remembered that the militarily defeated or colonised autochthons have no reason to like the states that usurped their sovereignty or to recognise their legitimacy; and that they object politically to the archaeology of the occupying power as just another form of cognitive colonialism (see below).

## Archaeology's curatorial ambitions

One of the most fundamental principles of the academy is the concept of academic freedom: the notion of an inalienable freedom of enquiry of researchers. The exact meaning of this concept differs somewhat in various countries, but the general idea concerns the liberty of thought, teaching and debate. Curiously, archaeology is a discipline that at the same time both embraces this principle and rejects it. This is one of the strangest contradictions in any academic pursuit.

The cognitive and intellectual colonisation of indigenous societies, be they extant or extinct, is based on the presumption that an academic endeavour has the right to investigate or study any subject matter, irrespective of whether this may be injurious to some subjects. In the Western intellectual tradition as it has evolved, the right to investigate is held to be a fundamental principle of academic research, and is also embraced by archaeologists. They sometimes find themselves in the courts, opposed by indigenous groups who have different views (cf. Barkan and Bush 2002). A classical example was the wrangling over scientifically unimportant midden contents from Tasmania, detailed below. It was based on two specious notions: that archaeology is a science, which we have seen it is not, and that a science has unfettered 'rights to know'.

The monumental self-contradiction occurs when archaeology at the same time rejects the principle of academic freedom. Most obviously, it restricts the right to excavate a site to specific accredited members of the discipline, irrespective of their relative competence. Unless one has a university degree in archaeology and is defined as a 'professional', one is not allowed to conduct excavation of an archaeological site. This is, first of all, an absurd rule, because there is hardly any piece of ground in the world that could not reasonably be described as an 'archaeological site'. Secondly, it assumes that only an archaeologist who is paid can have the required competence; indeed, the assumption appears to be that payment (defining professional status) is a measure of competence. This is another absurdity, it favours the politically adept, and negates the principle of academic freedom: no astronomer would ban amateurs from studying the stars, no palaeontologist excludes knowledgeable amateur palaeontologists, and no chemist considers banning others from working with chemicals. Nevertheless, archaeology manages to enforce this principle to some extent, by effectively banning amateur archaeologists from working, irrespective of their level of competence. This is then another imposition of political power over academic freedom. Moreover, in this the discipline ignores that practically all really important discoveries of archaeology, be they practical or theoretical, were contributed by people other than 'professional' archaeologists — who in fact contributed little of consequence to the discipline (but numerous academic follies; for some of the more prominent examples, see Chapters 4 and 5). Perhaps even more importantly, in assessing the epistemology of archaeology, it is also ignored that the errors of archaeology were almost always corrected by outsiders, as we shall see.

So there are several reasons why mainstream archaeology might profitably draw on resources external to it, but as a discipline it rejects this idea categorically. Instead it tends to surround itself with protective and restrictive structures and practices. It limits access to its data and collections to its accredited practitioners by a variety of means. For instance, many archaeological reports and results are privately owned, typically by the corporate masters of archaeological consultants. Or they may be held in supposedly public agencies of the state that can in various ways frustrate 'outsiders' from gaining access. In cases where immensely powerful corporations are involved in the control and exploitation of valuable economic resources, related archaeological work tends to be secretive and avoids public attention as best it can (Laurie 2006). Again we see the political role of archaeology in the service of the existing power hegemony, and we see the corruption of principles of academic freedom.

However, the principal opponents to the political powers of archaeology are the indigenous peoples of the world, because they object to the usurpation of their culture and history by dominant hegemonies. Around the world there have been many political confrontations between archaeologists and traditional peoples, in all continents other than Antarctica, including in Europe (e.g. with the Saami), over a variety of issues, including the curatorial aspirations of archaeologists. Archaeologists may find it surprising that an argument can be made that their reason for often opposing indigenous aspirations is because indigenes are actual competitors of archaeologists. But that is the case: indigenes are the true experts on their pasts (possessing emic knowledge of it), and they compete with archaeologists for possession of material culture.

For instance in Australia, Aboriginal people have frequently needed recourse to the law to gain both respect for and control of their cultural heritage. Notable early examples are the cases of *Foster v Mountford* in 1976 (29 FLR 233), *Pitjantjatjara Council v Lowe and Bender* (unreported, 25 March 1982), *Berg v University of Melbourne* (unreported, 18 June 1984) and *Berg v Museum of Victoria* (1984; VR 613). The material in question is often skeletal, but may also include artefactual material, even midden remains, as in the following case. The archaeology department of La Trobe University in Melbourne, ostensibly basing its claim on the perceived 'academic rights of science', refused to return some 130,000 items (generally stone artefacts and faunal remains) to the Aborigines of Tasmania, represented by the Tasmanian Aboriginal Land Council (TALC). In July 1995, after extensive unproductive negotiations with the Tasmanian state government (a government of racial discrimination, as noted above), TALC initiated litigation in the Federal Court. Professor Jim Allen from the university opposed the return of the material after it had been in storage for several years without being subjected to detailed study. The five permits for temporary possession had expired, no new permit was granted, and the issue was not about some material of great scientific importance, but about the principle that the archaeologists exercised control over their finds in perpetuity. The court directed the university to hand the material over and it was repatriated to Tasmania.

This led to hysterical reactions from some archaeologists, such as an article, 'The death of archaeology' (Murray 1995), and the use of the argument that archaeologists gave Aborigines "confidence and pride in their prehistoric achievements". This implied the notion that Aborigines ought to be grateful to archaeologists, not oppose them, rather as if archaeology was conducted for the benefit of the people being studied. Adverse press comments railed against the waste of taxpayer's money, first in amassing tons of materials that nobody ever studied, and then in expensive but equally unnecessary litigation. Archaeology, after all, is being conducted for a variety of reasons, including consolidating the power of the state and the academy, and the advancement of individual academics within the latter. Idealistic support of indigenous causes is not its primary motivation.

More recently another Australian archaeology professor prompted a related public debate. Iain Davidson, from the University of New England, who infamously acted as an apologist for a large company engaged in the destruction of rock art (Bednarik 2006b), also bought into the polemics concerning the property rights of members of his profession over archaeological finds. In connection with the highly publicised dispute between Indonesian and Australian researchers over possession and interpretation of the newly-discovered bones of humans from Flores (called *Homo floresiensis* and dubbed 'Hobbit', see Chapter 5), Davidson had stated in a newspaper that "it was the sole right of the team finding the remains to decide who had access to them". In February 2005 I reminded him that the archaeologists were servants of the state and had no such property rights, that this restricted access to researchers of opposing views and that the highly controversial finds were in any case Indonesian. I also asked him to clarify for how long he thought the finders should have exclusive possession of the finds. He nominated the duration of the funding period as a minimum, which I rejected on both ethical and epistemological grounds. Most importantly, it refused academic freedom to those with opposing views.

The perverse demands of some archaeologists that they should have unfettered control over finds or sites, excluding even their own colleagues if they disagreed with them on interpretation, are only one of several expressions of curatorial ambitions. These are clearly unscholarly, restrictive and without legitimate basis; they express a political desire for greater power. It appears that they also may be due to a misunderstanding: some archaeologists seem to subscribe to notions of the rights of independent scholars, such as those of the 19th century. These 'gentlemen researchers' were of course amateurs, in modern terminology, and may have had some personal though debatable claims over materials found. So the modern archaeological servant of the state and of the public seems to assume that he has similar rights, while at the same time brandishing his 'professionalism' (i.e. his dependency on the state) at every opportunity when opposing independent researchers. He seems to want it both ways: when the question of intellectual ownership arises, he tends to forget that he was paid, and thus owns neither the objects, sites, reports, or any intellectual property he produced while in the employ of someone else. Much the same, conversely, applies right through the academy, when it comes to intellectual property rights: the person who produces something for payment should not own the result of the labour; his master does. Thus the *independent* researcher or amateur owns his work; the paid academic does not, he or she is a *dependent* researcher. Which is why in some countries, paid public servants may not benefit financially from work they produce incidental to their employ, e.g. books they write.

Another self-contradiction of archaeologists is when they assume that, in many cases, the societies they investigate are extinct, therefore have no objection to being investigated. But on the other hand, the very reason for developing archaeologies in the first place was to study the precursors of present nation-states by investigating their origins through 'extinct' societies. If this were the case, then all previous societies would have cultural heirs, and these can object to being subjected to the revisionist histories archaeology produces. In the same way as Aborigines or San or Inuit have the moral right to object to the attention of archaeologists (or anthropologists), so should every citizen.

## Archaeology and 'the other'

All political and ethnic groups of humans define themselves, in part at least, by contrasting themselves with others, and traditionally this is by viewing the others in condescending ways. These are inevitably regarded as inferior, in any culture, nation, or ethnic group. Indeed, diminishing, dehumanising or even demonising those 'others' often facilitates group cohesion. This tendency is in a general sense related to the inherent fear of the non-conforming we find in many social constructs: the saints, criminals, mental patients, geniuses, foreigners, heretics, the rude and obnoxious, the truly assertive are all perceived as threats to the social reality and fabric of groups. (For instance many archaeologists reading my words here will find them threatening, objectionable and offensive, because they treat their discipline as a social construct rather than a science; a science would be capable of discussing *any* issue in a detached intellectual format.)

One of the most pernicious political aspects of archaeology is the effect of its particular relationship with its subjects: it seeks to understand our past by necessarily contrasting 'us' with those perceived to be more primitive than 'we, the civilised'. It has to be inherently concerned, in many cases, with 'the other': peoples or 'races' other than one's own — the latter being most often White and Western. Archaeology has no choice but to contemplate the 'cultures' of 'the others' from a perspective of considering its own methods of acquiring knowledge superior to those of the societies it professes to study. Its pursuit would make little sense if practitioners were to concede that these other cultures were superior: judging the workings of a superior system and expecting much validity would seem futile. Since all cultures studied by archaeology are earlier or technologically less advanced, it is readily assumed that they are simpler, that they can effectively be analysed by members of the technologically dominant society. Technology, then, is the measure of sophistication in this relationship.

Needless to say, this is a misconception. Technological superiority is no proof of cultural, moral, ethical, cognitive, or even intellectual superiority. Indeed, as pointed out in Chapter 1, Western civilisation views knowledge and power as equivalent, instead of linking knowledge and morality. This is not only undeniably true (we even value the aphorism 'Knowledge is power', a fair indication of the degree of Western intellectual corruption), as Western civilisation totters towards its own gradual destruction, because it ignores this principle; it also demonstrates its moral and ethical inferiority. These issues were popularly explored in the 1968 film *Planet of the apes*, featuring a post-apocalyptic Earth culturally dominated by apes, holding up a mirror to our academy. The ape academy scoffed at the preposterous notion that humans were once more advanced than apes, and when its director, Dr Zaius, realised that he had been wrong he destroyed the evidence deliberately to maintain his dogma (Figure 8). Equivalent occurrences in archaeology will be discussed in these pages.

Nevertheless, academic archaeology takes it upon itself to define, judge, quantify, and describe the peoples of the past, those 'others' of history. In purely epistemological terms this means that we, who have great difficulty comprehending the reality we ourselves created (Searle 1995), dare to attempt defining the reality constructs of the 'others', our 'subjects'. While this is of course a legitimate academic pursuit, we must not expect it to yield finite truths and we must duly respect these 'others' (Heyd 2007). It is, nevertheless, always very difficult to prevail over the inherent ethnic partisanship of all archaeology and history (Wailes and Zoll 1995: 23). For instance, we need to avoid discriminating between those we identify as our ancestors and those we exclude from them. Another issue tends to creep into glorifying our own societies' pasts, a predilection of emphasising differences at the expense of the 'others'. Although most expressions of ethnic archaeology became 'politically incorrect' by the

*Figure 8. Scene from* Planet of the apes *(1968 version), in which Dr Zaius realises that the imprisoned Taylor is intelligent but destroys the evidence to preserve the ape dogma.*

middle of the 20th century, in part because of the defeat and subsequent erosion of fascist governments in Europe (in Spain and Portugal occurring well after World War 2), these currents survived in alternative dimensions. Most particularly, the handmaiden of both fascism and archaeology, religion (especially Christianity), has always had severe concerns about differentiating between the 'crown of creation' and other beings. In past centuries religion had been able to deal with the 'others' by defining them as unworthy heathens, but through recent centuries much of the power of this idea had been eroded by more enlightened ideologies. These also marked the decline of religious power, the gradual diminishment of slavery and the rise of the idea that women should have political rights. The final shock came in the mid-19th century, with the European discovery of a fundamental wisdom long understood by all the 'heathens' of the world: that humans are biologically animals, that they descend from other animals. This threatened the very foundations of Christianity, which was based on the notion that humans (Aryans, especially) had been created in God's own image. If noble Europeans were the descendents of 'more primitive' folk, and ultimately of apes, the inevitable question was: at what precise point in time, and for what specific reasons, did they suddenly become humans, capable of entering heaven?

One of the several ways Christianity fought enlightenment was to encourage clerics to take up 'prehistoric' studies, enabling them to influence the new discipline's direction and public impact. In trying to emphasise the division between humans and other animals, numerous devices were employed, ranging from tool making to self-awareness. Yet tools were used by countless species, and self-awareness can be assumed to exist, or has existed, in at least a dozen species (all known and unknown hominins, in chimps and bonobos, and very probably in others still; cf. Bednarik 2011), and its level in the contemporary human is debatable (most of what we tend to regard as self-awareness relates to muddled religious, ideological, ontological, academic and cognitive a prioris, and to simple biological equipment, such as proprioceptors). Moreover, today thousands of archaeologists practice Biblical archaeology, apparently unaware that the very concept is an oxymoron. The nebulous purpose of this field, which we visited already in Chapter 1, is not only the confirmation of the contents of one single book, but also a watering down of the more 'strident' strains of fervency in the discipline. But it is no longer socially acceptable to discriminate against extant societies, and missionary activities among them have been curtailed significantly, or have been slanted towards humanitarian aims. Archaeology now confronts the 'other' in new contexts on behalf of laudable Christian values. No longer is it acceptable to depict non-Christian societies as inferior, and the last-ditch endeavour on behalf of religion is the new frontier of defending the special status of humans in the Pleistocene context. This is directly related to the fundamental need of religion to find an arbitrary demarcation between those capable of salvation, and those too primitive to qualify.

An epistemologically similar aspect of Pleistocene archaeology derives from the position that absence of evidence equals evidence of absence. Some Pleistocene archaeologists argue that if we cannot perceive any evidence for communication then that capability does not exist archaeologically. Some even extend this to non-archaeological topics, contending that animal communication exists only when it is perceived by humans (Davidson 1992). This ultra-empiricism expresses an ontology in which current human knowledge alone determines how things really are in the world, and things do not exist until humans become aware of them. It is an underlying ideology in archaeology; e.g. we have no evidence of human occupation in North America predating Clovis, hence that continent was not occupied earlier. Irrespective of its incompatibility with South American evidence, it is clear that this pronouncement rests entirely on negative evidence — on the absence of evidence. Such conservatism may seem commendable, but it is applied in a completely random fashion. We have no hard evidence prior to 6000 BP that hominins possessed soft tissue or hair similar to our own, but archaeologists assume that they did; we have no material evidence of navigation prior to 9500 BP and yet it is generally accepted that vessels were used by Middle Palaeolithic people. In numerous cases, archaeologists do not require hard evidence to accept the existence of a phenomenon, in others they require hard evidence in the form of large numbers of incidence (e.g. for symbolism). Not only does this indicate the exercise of unexplained double standards in the demands for evidence, the division between those phenomena requiring hard evidence and those not requiring it seems entirely random, in the sense that there appear to be no rules determining these categories. To take the above example of animal communication: no ethologist would dream of using human capability of detecting animal communication as a measure of such phenomena, because ethologists, as scientists, begin with the assumption that we do not know how things really are in the world (if we did we would not need to find out). But when the student of human ethology (and 'prehistoric archaeology' is just an unscientific name for human ethology) considers the communication ability of hominins, a totally different epistemology takes over: we assume that communication is absent unless there is glaringly hard evidence in its favour (e.g. iconic pictures of things, according to Davidson and Noble 1989).

Today the frontier of religion-inspired ideology in archaeology finds its finest expression in the 'African Eve' model, which stipulates that all 'modern' humans descend from a single female, and that only her progeny has survived. All other humans have now become the new 'others' (Figure 9), which has the considerable benefit that they cannot defend their case and, apart from a few researchers who are too much concerned about veracity, nobody objects much to this version of the human past. Nobody seems to mind that it is a cynical exercise of ethnic archaeology, which has been driven out of more recent periods, justifying the definition of some 'other' as inferior and primitive,

*Figure 9. The author in conference with Mr Neander, who is much offended by what archaeologists have said about him.*

as the antithesis of our glorious ancestors. The African Eve or Replacement Model is the current archaeological myth of how the world's brutish forces were defeated. Chris Stringer and his coterie of supporters would indignantly reject the notion that he might have been influenced by Christian values, but the thought-patterns of Europeans, and many others, bear the indelible imprint of their religion — however dedicated they may individually be as atheists or agnostics. I cannot escape the fact that my culture is largely based on Christian values (e.g. the calendar I am forced to use; just as Stringer's Christian name is, well, Christian), which have in the past been the basis of slavery, fascism, pogroms, crusades and inquisitions. To suggest that any of us can entirely escape the cognitive conditioning of the dominant culture (and the religion it is based on, no matter how non-religious we may be as individuals) is foolhardy.

But we could try.

## Neo-colonialist effects of contemporary archaeology

In discussing the purported 'world archaeology' it is requisite to consider the effects of such global developments on the research traditions of non-Western countries: the socialist or former socialist countries, and those of the developing world. It may at first glance appear that such structures would benefit developing countries, but on closer examination and after considering the existing practices and research philosophies, the issue is considerably more intricate. Local research traditions already disadvantaged

*Figure 10. Professor Bruce Trigger, 1937–2006.*

may be affected even more adversely, and there is the complex issue of the political dimensions of archaeology.

The late Bruce Trigger (Figure 10), we have noted (e.g. in the conclusion of Chapter 2), perceived only three types of archaeology: nationalist, colonialist and imperialist. We have visited some of the more obvious manifestations of these aspects, but some others are perhaps less apparent. There are subtle biases in favour of the wealthy countries, which tend to facilitate an inconspicuous neocolonialism through monopolisation. For instance there are distinct preferences for methods that are sophisticated, complex, expensive and monopoly-forming. If there is a choice between two apparently similarly reliable methods, the one that is much cheaper and requires little expertise will quite likely be rejected. It is not the case that researchers in wealthy countries wish to exclude those from poorer countries; they are most welcome to solicit assistance from institutes in those richer countries. But their dependence on this help from Western countries may be induced by this preference for methodologies facilitating the establishment of scientific monopolies.

Similarly, prominent and outstanding archaeologists tend to publish their best work in the most prestigious journals, which today are mostly in the English language. Do their choices reflect a desire to educate or inform, or are they influenced by ambition and self-advancement? Scholars in developing countries may not be in an economic position to subscribe to these prestigious (and expensive) Western journals, they may rarely get to see such work. This is in addition to the 'brain drain' that occurs in any case in the poorer countries. Another problem with how archaeology is disseminated globally concerns the manner in which audiences and themes are constructed. Both are part of the framework of those who write, deriving from a relatively narrow band within predominantly Western academia, and those who read. This is a self-feeding cycle, considering that the pedagogical discourses depend on these same editorial policies.

Academic disciplines may even reject methods that are perceived to pose a threat to their control. When in 1980 I introduced the first successful attempt to date rock art via a direct, non-archaeological method my paper was rejected by five archaeologist referees, essentially on the basis that they doubted the validity of the geomorphological method employed. This was even though they admitted never having heard of it, and despite the references to previous similar work in speleology. These references, admittedly, were all to German and French papers, which points to another problem in Anglophone archaeology we have visited before.

Since then, the development of non-archaeological dating methods for rock art has helped rock art studies to break away from traditional archaeology. Ten years later, I developed another dating method, this time with the implicit purpose of providing a cheap, easily applied and almost universally available technique that would be well within the reach of most specialists in developing countries. It was refereed by two leading protagonists of highly sophisticated rock art dating methods. Both rejected the paper categorically without giving any credible technical reason. They found the idea that one might date a petroglyph without the use of millions of dollars' worth of hardware, and date it perhaps even better and more reliably, so unpalatable that they did not even want the method discussed or considered.

Researchers in developing countries need to understand these dynamics, and that their indebtedness to their Western 'sponsors' may not be quite as justified as it seems: some of it may have been induced by restrictive practices that may be illegal in commerce, but nevertheless flourish in academia.

These may be viewed as relatively peripheral issues, and it is admittedly true that the various fundamental epistemological issues discussed above are more effective in presenting us with archaeological models marked by systematic deficiencies. The political roles of the discipline, its categorical lack of a scientific universal theory, its tolerance of pathological practices, its several inherent self-contradictions and its cognitive colonisation of other cultures and societies all conspire to render the discipline an exercise in neo-colonialism — even if individual practitioners bravely oppose such currents and resist their influence. In the end archaeology has long become an institution of the state, an integral part of its technocracy. Those who could have prevented this from happening, the independent archaeological scholars, have been gradually marginalised over the entire 20th century, and today are perceived as troublesome interlopers that need to be opposed at every opportunity. Today, the academic support indigenous peoples enjoy from this discipline often derives from its dissenters, its heretics and outsiders. It is truly a sad state of affairs, and one that all the media propaganda the state's machinery churns out cannot paint over.

## REFERENCES

Barkan, E. and R. Bush (eds) 2002. *Claiming the stones, naming the bones: cultural property and the negotiation of national and ethnic identity*. Getty Research Institute, Los Angeles.

Bates, D. 1944. *The passing of the Aborigines*. John Murray, London.

Bednarik, R. G. 2006a. Rock art protection in Tasmania. *Rock Art Research* 23: 282.

Bednarik, R. G. 2006b. The 'Assistant Undertaker' of Dampier rock art. *AURA Newsletter* 23: 10–11.

Bednarik, R. G. 2011. *The human condition*. Springer, New York.

Berger, P. L. and T. Luckmann 1966. *The social construction of reality: a treatise in the sociology of knowledge.* Anchor Books, Garden City, NY.

Chaloner, T. 2004. The Aboriginal Heritage Act 1972: a clash of two cultures; a conflict between two laws. A parliamentary internship, a co-operative arrangement between The Hon. Robin Chapple MLC of the WA State Parliament and Murdoch University.

Davidson, I. 1992. There's no art — To find the mind's construction — In offence. *Cambridge Archaeological Journal*, 2: 52–57.

Davidson, I. and W. Noble 1989. The archaeology of perception: traces of depiction and language. *Current Anthropology* 30(2): 125–155.

Escobar, A. 1991. Anthropology and the development encounter: the making and marketing of development anthropology. *American Ethnologist* 18(4): 658–682.

Heyd, T. 2007. Cross-cultural contact, etiquette and rock art. *Rock Art Research* 24: 191–197.

Howsbawm, E. and T. Ranger (eds) 1983. *The invention of tradition.* Cambridge University Press, Cambridge.

Houtman, G. 2006. Double or quits. *Anthropology Today* 22(6): 1–3.

Houtman, G. 2007. Response to Laura McNamara. *Anthropology Today* 23(2): 21.

Hulley, C. E. 1996. *Dreamtime Moon.* Reed Books, Chatswood,

Johnson, D. 1998. *Night skies of Aboriginal Australia: a noctuary.* University of Sydney Press, Sydney.

Laurie, V. 2006. Burrup treasure is history in the taking. *The Australian*, 31 October, p. 10.

Leone, M. and P. Potter 1992. Legitimation and the classification of archaeological sites. *American Antiquity* 57: 137–145.

McNamara, L. A. 2007. Culture, critique and credibility. *Anthropology Today* 23(2): 20–21.

Moore, P. 1999. Anthropological practice and Aboriginal heritage: a case study from Western Australia. In S. Toussaint and J. Taylor (eds), *Applied anthropology in Australasia*, pp. 229–254. University of Western Australia Press, Nedlands.

Murray, T. 1995. The death of archaeology. *Campus Review* 5(34), 31 August.

Pinker, S. 2002. *The blank slate. The modern denial of human nature.* Penguin Putnam.

Price, D. 2000. Anthropologists as spies. *The Nation* 271(16): 24–27.

Price, D. H. 2005. America the ambivalent: quietly selling anthropology to the CIA. *Anthropology Today* 21(5): 1–2.

Ritter, D. 2003. Trashing heritage. *Studies in Western Australian History* 23: 195–208.

Sapir, E. 1929. The status of linguistics as science. *Language* 5: 207–214.

Searle, J. R. 1995. *The construction of social reality.* Allen Lane, London.

Silberman, N. A. 1989. *Between past and present: archaeology, ideology and nationalism in the modern Middle East.* Henry Holt, New York.

Sims, P. C. 2006. Rock art vandalism in Tasmania. *Rock Art Research* 23: 119–122.

Wailes, B. and A. L. Zoll 1995. Civilization, barbarism, and nationalism in European archaeology. In P. L. Kohl and C. Fawcett (eds), *Nationalism, politics, and the practice of archaeology*, pp. 21–38. Cambridge University Press, Cambridge.

Warner, W. L. 1937. *A black civilization: a social study of an Australian tribe.* Harper, New York.

Wells, A. E. 1964. *Skies of Arnhem Land.* Angus and Robertson, Sydney.

# 4

# MILESTONES OF PLEISTOCENE ARCHAEOLOGY

## Heretics in Pleistocene archaeology

The most characteristic feature of archaeology is not that it deals with the past; many disciplines do that, including palaeontology, palynology, geology, or astronomy. Nor is it that archaeology conjures up images of mystery and adventure; most disciplines can do that. Nor that it often deals with interesting remote places and countries. Previous chapters suggest that archaeology as we know it is more readily characterised by a collection of rather negative factors. There is the uneasy relationship it has developed with its two principal client groups: the interested public, which it relegates to the status of spectators, 'cult archaeologists', 'folk archaeologists', or amateurs (in the derogatory sense); and the political structures in the struggle of indigenous people around the world who in many cases object to archaeological practices. The tensions are in both cases not only due to excessive curatorial desires of the discipline's practitioners and the political agendas they serve, but also due to archaeologists' misunderstandings of the role and capabilities of archaeology, and the ethical fact that archaeological property does not belong to state-appointed 'experts'. They speak neither for indigenous peoples nor for science; in fact their curatorial demands conflict with principles of academic freedom as well as the aspirations of cultural autonomy of autochthon peoples.

But by far the most characteristic feature of establishment archaeology is its treatment of heretics and iconoclasts: the people who disagree with its established dogmas. It is not so much that heretics may not also be rejected in other disciplines, but in other fields of human endeavour there is a tendency to learn from mistakes made in rejecting heretical views. This is not evident in archaeology. That discipline has had to deal with significant dissent for its entire history, and that history provides evidence that archaeology as an academic discipline has learnt nothing from these encounters — or from the great embarrassments they have led to. Today it uses precisely the same strategies of silencing these people it used almost 200 years ago. This is what I seek to demonstrate here. To see this, and to understand this extreme conservatism it is necessary to examine some case histories.

There are hundreds of cases where non-archaeologists offered important ideas, data or finds to the discipline, only to be rejected — and usually so with great displays of indignation — and where it was subsequently found that the finds were authentic, the data valid or the ideas extremely important in developing archaeology. Often this realisation that the discipline had made a great error in rejecting such outsiders only came after the death of the heretical scholar. The most interesting aspect of all of this is that it did not prevent the discipline from repeating the same treatment of iconoclasts, decade after decade, right up to the present time. This is happening unabatedly today, and the pattern of reflex-like rejection, and much later grudging acceptance is by far the most characteristic aspect of archaeology as a reactionary discipline. In this sense, orthodox archaeology is again reminiscent of some religions.

Archaeology does not appreciate being compared with religion, and yet it has strong dogmas, it 'crucifies' heretics, it is a belief system, and its history is inter-woven with specific religions. 'Biblical archaeology' remains a valid subject at Western universities, and the absence of Koranic or other religiously motivated forms of archaeology at these same universities belies their claims of objective scholarship. Such institutions are dedicated to intellectually corrupt forms of scholarship, including the misuse of scientific techniques to demonstrate the validity of what the neurosciences call magical thinking. This state of affairs, we have noted, is reflected in modern archaeology by such aspects as the 'African Eve' model, which recently enjoyed great popularity.

In order to examine possible historical patterns in the treatment of archaeological heretics it is most instructive to consider the most celebrated cases in the history of the discipline. It needs to be appreciated that no *truly* important archaeological discovery was ever made by a professional archaeologist. The most important finds made in Pleistocene and Pliocene archaeology are perhaps the discoveries of man's antiquity, of fossil man, of Palaeolithic cave art, of *Homo erectus*, and of *Australopithecus*. They were all made by non-archaeologists, and they were always rejected by archaeologists — in all cases without seriously attempting to examine the evidence fairly.

## The discovery of humanity's great antiquity

One of the first to recognise the immense antiquity of humanity was Jacques Boucher de Crèvecœur de Perthes (1788–1868). He was a French customs official who for decades in his spare time collected Palaeolithic stone tools, such as hand-axes, in the gravels of the River Somme (Figure 11). In examining finds made by a local medical doctor, Casimir Picard, in the 1820s he recognised these as the handiwork of human beings, and finding

*Figure 11. Jacques Boucher de Crèvecœur de Perthes.*

them together with the bones of extinct animals he realised that humans must have lived in France during the Pleistocene (the Diluvium, as it was then called). A year after having been seconded to Abbeville in 1825 he began to collect stone implements and he soon became a regular visitor to the region's quarries, canal diggings and gravel pits. About 1832 he commenced systematic excavations, amassing a large collection of flint tools and other material, and by 1838 he presented his theory and evidence to the Société d'Émulation. Undaunted by the scepticism that greeted him, he did the same in the following year before the Paris Institute, where his ideas and finds were thoroughly rejected. He then published his work in five volumes entitled *On the Creation*, again finding it rejected by the 'experts'.

Unfortunately this man also had some other eccentric ideas: he thought that women should have rights, he suggested the raising of the living standards of the working classes, and he advocated universal peace. In short, it was easy to see that the Boucher de Perthes was just a crank. Nevertheless, he proved to be a very persistent crank, and during the late 1840s, Scandinavian archaeologists had begun to turn the tide in his favour, with their Three Age system (Stone, Bronze and Iron Ages). By that time he had become so confident that he claimed humans existed many thousands, even hundreds of thousands of years ago (Boucher de Perthes 1846). This he deduced from the geological age of the strata he had excavated, and he quite correctly pointed out that most tools then might have been made from wood, but that in order to work wood the use of a harder material, flint, was essential.

Of particular interest here is Boucher de Perthes' own reaction to the consistent rejection he experienced over some decades: "They employed against me a weapon more potent than objections, than criticism, than satire or even persecution — the weapon of disdain. They did not discuss my facts; they did not even take the trouble to deny them. They disregarded them." This is important, because the same weapon is still widely employed in contemporary archaeology, and in precisely the same way.

The final denouncement came in 1858, at a French archaeology congress, which issued a unanimous declaration according to which all of de Perthes' stone tools from Abbeville were "a worthless collection of randomly picked up pebbles" (Figure 12). The archaeologists, who had long objected to the disciplinary trespass of this troublesome amateur, had realised that Boucher de Perthes was not going to give up easily, and that he was even gaining a little support, especially from two more amateur archaeologists, another doctor and a geologist: Marcel-Jérôme Rigollot and Edmond Hébert. So at long last they decided to act decisively against these cranks. This turned out to be a great and rather untimely embarrassment, because in the following year, two British geologists, who had quietly worked away testing Boucher de Perthes' propositions (which is precisely what good scientists do), published their findings (Prestwich 1859), confirming that he had been right all along and the 'worthless pebbles' were the tools of 'Diluvial man'. Hugh Falconer and

*Figure 12. Some of the "worthless pebbles", Musée Boucher-de-Perthes, Abbeville.*

Joseph Prestwich had themselves become aware that he might be right after taking part in the 1858 supervised excavation of Windmill Hill Cave, Brixon, by another autodidact, William Pengelly (1812–1894). Pengelly, a self-taught geologist, had earlier excavated in Kents Cavern where, continuing the work of John MacEnery, he had found extinct animal remains together with flint tools, and repeated Boucher de Perthes' claims in the late 1840s.

Over the following years, in the wake of Darwin's (1859) and Lyell's (1863) influential publications, public opinion swung around sharply, and it is important to note that archaeological opinion followed suit. As we will see later, this is the usual pattern: archaeology follows public opinion, it is the most populist discipline, always ingratiating itself with the public. Boucher de Perthes lived to see his perseverance of half a lifetime vindicated because he addressed the public. Those heretics of archaeology who failed in this were not so fortunate to witness their exoneration.

## The discovery of fossil man

In the very year of another amateur's book, Darwin's 1859 seminal volume and Prestwich's substantiation of de Perthes' claims, an article by an unknown author appeared in a German journal. Johann Carl Fuhlrott (1803–1877) was a schoolteacher, and he presented a paper about what he claimed to be skeletal remains of a pre-Historic human being that probably was of the Diluvial period (Figure 13). But instead of welcoming the opportunity of publishing the first report ever of fossil

*Figure 13. Johann Carl Fuhlrott.*

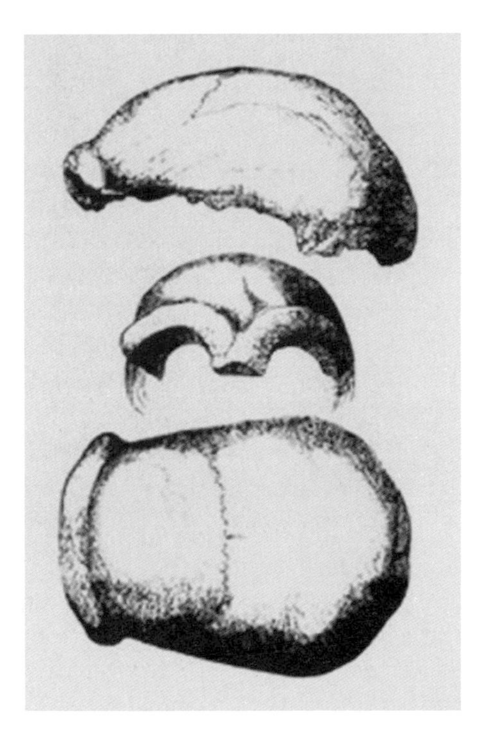

*Figure 14. The Neander valley finds of 1856.*

man the journal explicitly rejected the interpretation of the find, publishing with it a footnote expressing its disagreement with Fuhlrott's opinion (Fuhlrott 1859).

The bones had been excavated in August 1856 by two workers in a limestone quarry in the Neander valley in Germany (Figure 14). In removing the sediment fill of the Kleine Feldhofer Cave, they threw the bones on the slope of waste material, where an owner of the quarry noticed them. Thinking that they were cave bear bones, he collected the larger ones and gave them to Fuhlrott, whose work as a naturalist was known locally. Fuhlrott realised that the individual represented by the bones differed significantly from modern humans, and he also recognised that the clay deposit in the cave seemed to be of the Ice Ages. In 1857 he presented these findings to a conference in Bonn where they were rejected, except for the anatomist Hermann Schaaffhausen, who had examined the bones and tentatively agreed with the teacher from Eberfeld.

In 1860, the founder of geology, Charles Lyell, visited Fuhlrott and the Neander valley, taking a plaster cast of the cranium, and Thomas Henry Huxley, Darwin's most outspoken supporter and yet another amateur, commented that it was the most ape-like human skull he had ever seen. His view was shared by Irishman William King, and one would have thought Fuhlrott's views would have been accepted within a few years, particularly in view of the rise of Darwinism at that time. Far from it: over the following years, the remains were variously attributed to a Mongolian Cossack, a Celt, a Dutchman, a Friesian, and an idiot. The bone architecture was attributed to various bone diseases, the curved leg bones to a life of riding horses. The raging controversy was 'resolved' in 1872 when Germany's foremost expert, Professor Rudolf Virchow (1821–1902), entered the fray at long last (Figure 15). As the president of the Society for Anthropology, Ethnology and Prehistory, and an anatomist of great prestige, he determined the health history of the individual since his childhood from the bones available, in a classical demonstration of the value of deductive diagnosis. His authoritative rejection of Fuhlrott's interpretation seemed to be decisive and public perception was guided by it.

*Figure 15. Professor Rudolf Virchow.*

That would have been the end of the story for some time if it had not been for the British Darwinists. Interestingly, Virchow was quite supportive of the evolution theory at the time, even though he defined it as not adequately supported by empirical data (while, of course, rejecting the empirical data). However, his position over the following years hardened, and when odd-looking human mandibles were found in two other caves (La Naulette in France and Šipka Cave in Moravia) he rejected them as being typical of a 'race'. By 1877 he warned against Ernst Heinrich Haeckel's proposal to teach evolution theory in schools. He felt it would "dispossess" the churches, and he became alarmed that socialists had adopted evolutionism into their political agenda. Virchow was politically active, and one of the founders of a political party, the German Progressive Party. He grew progressively concerned about the implications of political Darwinism and eventually also spoke out against biological Darwinism as "limiting academic freedom". Thus the position of Neanderthal man became subordinated to other agendas, such as those of academic prestige and monopolisation, political issues and tribal warfare among the tribes of academia. Fuhlrott had become irrelevant, a mere footnote in the discipline's history.

However, in 1886, after thirty years, he was finally vindicated. An excavation in a cave at Spy, near Namur, Belgium, produced two substantially complete skeletons of humans, found together with numerous stone tools and the bones of extinct animal species. The characteristics of the human bones matched those of the Neanderthal find, and the theory of congenital bone deformation collapsed. By now it was widely accepted outside of Germany that these were representatives of fossil humans of the Pleistocene, and yet, in Germany it took another fifteen years before a detailed favourable study of the Neander valley remains was published. It was in fact a student of Virchow, Gustav Schwalbe (1844–1916), who published a reassessment of the original Neander valley skeleton, defining it as the species *Homo neanderthalensis* (Schwalbe 1901).

## The discovery of Pleistocene art

The existence of Palaeolithic cave art was long known, probably always since the Pleistocene. For instance we know that in 1458 Pope Calixtus III decreed that the religious ceremonies held in "the Spanish cave with the horse pictures"

*Figure 16. Don Marcelino Santiago Tomás Sanz de Sautuola.*

had to cease. We cannot know which cave he referred to, but it was almost certainly a cave with Palaeolithic art. And there are even earlier inscriptions found with cave art in some sites, and innumerable more recent ones in such caves as Niaux and Rouffignac. However, while many people of the ten thousand years after the Pleistocene were perfectly familiar with the ancient art, nobody had told the archaeologists about it. This factor should turn out to be of serious consequences for Don Marcelino Santiago Tomás Sanz de Sautuola (1831–1888), a nobleman of the Santander region in northern Spain (Figure 16). In due course it would destroy his life. A landowner of wealth and diverse interests, he had been one of the first to introduce eucalypts in Spain, he had a magnificent library and a good knowledge of the region's geology and ancient sites. In fact he was appointed vice-president of the monuments commission of his district in 1872, just seven years before the fateful discovery that would ruin him.

The story begins in 1868, when a hunter, Modesto Cubillas, lost his dog on Altamira, a limestone hill on de Sautuola's property. It had climbed into a cave and found itself unable to come out. The hunter opened up a hole and found a large cavern. This was mentioned to the land's owner years later, in 1875, who decided to explore the cave. He found a large quantity of split bone upon digging in its floor deposit, some of which he took to show a geologist friend, Juan Vilanova y Piera at Madrid University. Vilanova recognised the bones as being from extinct species, and that they had been fractured by humans. In 1878 de Sautuola visited the World Exhibition in Paris, which included an exhibit of stone tools and bones recently excavated in caves of the French Périgord. De Sautuola remembered his own cave and, in the spring of the following year, began in earnest to excavate part of the Altamira cave. Mixed with the bones of animals and oyster shells, he found the typical stone blades of the Magdalenian period in large quantity. Deeper in the cave, a complete skeleton of a cave bear was encountered, and the explorer also observed black markings on the cave wall, but gave them no further thought at that time. It was his 12-year-old daughter Maria, who, playing in the cave as he was digging, first noticed that there were animal pictures on the ceiling. This was in November 1879.

It was clear to de Sautuola at once that the incredible gallery of paintings of bison he now began to see was probably the work of the same people whose debris he was digging up, partly because he had already observed seashells full of paint pigment, and some of their debris occurred right on top of the floor deposit. He reported this incredible discovery immediately to Vilanova, who came to inspect the find. Upon examination of the floor sediments Vilanova agreed with his friend that the many paintings were ancient. He presented a lecture in Santander, the discovery made headlines across Spain, and King Alfonso XII visited the Altamira cave. In 1880 de Sautuola produced a publication, describing the paintings and the occupation evidence, but cautiously avoiding the claim that the two forms of evidence necessarily needed to belong to the same time. It was a sober treatise, entirely lacking in flamboyant claims. For the required illustrations he employed a destitute and dumb French painter he had befriended earlier, and this turned out to be a fatal mistake.

The publication was greeted with considerable disapproval, which soon built up to ridicule and anger. The discipline of archaeology and prehistory decided collectively that de Sautuola was either a charlatan, or at the very least he had been severely duped. At the International Congress of Anthropology and Prehistory in Lisbon, where the elite of Europe's 'prehistorians' gathered, Vilanova presented the discoveries in Altamira, strongly defending de Sautuola. One of the most influential French delegates, Professor Émile Cartailhac, walked out in disgust, and later roundly declared the paintings to be a fraud, without even bothering to see them. In fact all other experts refused to examine the site initially, and the French decided that the whole affair was a plot by Spanish Jesuits to undermine the credibility of pre-History as a discipline. Once again we see the struggle between the discipline and the Church made explicit. Eventually, a railways engineer who had a good knowledge of palaeontology, Édouard Harlé, was requested to examine de Sautuola's outrageous claims, and he promptly discovered the involvement of the dumb painter (who in the meantime had disappeared). No further investigation was needed, the case was clear enough to him.

Vilanova tried in vain to use his academic prestige to promote acceptance of the find; he was judged to have been the first to be duped by the charlatan of Altamira, and unable to concede that. De Sautuola, for his part, did not respond to the accusations. As a Spanish nobleman he felt that he could not enter into a discussion of whether he was honourable or not, but we know that he suffered greatly. He tried to present his case at a French conference in Algiers in 1882 and submitted a self-funded booklet to another conference, in Berlin, but both endeavours were ignored. Six years later he died at the age of fifty-seven, a broken and bitter man, in the full knowledge that he had made one of the greatest discoveries in the history of archaeology. He also knew that he had failed in effectively conveying this knowledge to a thoroughly hostile academic world. His death weighs heavily on archaeology, particularly as he was judged without trial — but all to no avail. As recently as 1996, this discipline has been responsible for the death of a researcher in Australia, Dr David Rindos. Being an archaeologist certainly is a health risk, and as one

Figure 17. Professor Émile Cartailhac.

practitioner, Paul Bahn, observed, it requires the hide of a rhinoceros.

A French schoolteacher, Léopold Chiron, had found engravings deep in the cave of Chabot already in 1878, and in 1890 in another site, Figuier. In 1883 Francois Daleau excavated engravings on a wall in Pair-non-Pair that had been covered by Pleistocene sediments. In 1895, a bison engraving was discovered in the French cave La Mouthe, and Emile Rivière, who had actually seen the Altamira paintings, found more rock art in La Mouthe, and four years later a Palaeolithic lamp. Thus the evidence in favour of Palaeolithic rock art mounted. In 1897 Cartailhac (Figure 17) still refused to publish the note of a new discovery of cave art, but in 1902 he published his famous '*Mea culpa d'un sceptique*', in which he grudgingly accepted that he had been monumentally wrong.

## The discovery of *Homo erectus*

During the last decade of the 19th century, another of the greatest discoveries of Pleistocene archaeology was just in the making. Eugène Dubois (1858–1941) was a Dutch physician who had been bitten by the archaeology bug as a young man (Figure 18). Ernst Haeckel, who had supported the initial identification of Neanderthal Man, predicted the existence of a 'missing link' on the fossil record, a creature that would be an intermediate form between extinct apes and human beings. He had given it the name *Pithecanthropus* in 1886, and in the absence of any fossil finds at the time, had sketched out what such a creature might look like. In 1870 Haeckel (Figure 19) had lectured in Holland, and Dubois set out quite deliberately to find the remains of *Pithecanthropus*, deciding that the region of today's Indonesia was the right place to look for them. Bearing in mind

Figure 18. Eugène Dubois.

the almost complete lack of knowledge about hominin evolution at the time, this was certainly a most audacious plan — and not one likely to succeed too readily. A passage in a book by the British naturalist Alfred Russell Wallace (Figure 20), who mentioned South-east Asian caves and apes in connection with the origins of humans, was Dubois' main clue, apart from the reasoning of both Wallace and Darwin that these origins were to be found in the tropics.

*Figure 19. Ernst Haeckel.*

He signed up as military physician to go to Sumatra in 1879, specifically intending to look for the 'missing link'. His investigations of caves in Sumatra led to no exciting finds, however, and after two years, following a bout of malaria, he was transferred to the drier Java. Here he secured limited support from the colonial administration, and the use of convict labour for major excavations. He began to dig at several sites, finding the sediments rich in fossils. In 1891, while excavating on the Solo River, he recovered first a tooth, then a cranium, which he judged to be of a large, human-like ape. In the following year, a thighbone was found, clearly of an upright walking primate, and a second molar. In 1893 Dubois claimed to have found an apelike hominin at Trinil, on the basis of these four specimens.

Even the telegraphed reports preceding Dubois' return to Holland created controversy, and it was claimed that he had combined a human femur with the cranium and teeth of an ape. One member of the Dutch Zoological Society asked, if we were to continue searching at the Trinil site and found a second left femur, would that indicate that *Pithecanthropus* had two left legs? Or that he had two different heads, if a human skull were found another 15 m further? Dubois had no idea what awaited him, and when a few weeks after his arrival he addressed a congress in Leiden, the assembled experts began forming opposing camps. Interestingly enough, these were largely divided along national lines, for instance, most Germans

*Figure 20. Alfred Russell Wallace.*

saw the creature as an ape with some human features, the British as a human with some simian features, while the Americans shared Dubois' view.

Over the following years, this debate raged without any agreement in sight. In 1928 Dubois recorded wryly that no less than fifteen different interpretations of his fossil finds had been proposed by as many scholars. He made great efforts to contribute to resolving the controversy, travelling widely, and learning the skills of a dentist, a photographer and a sculptor, creating a life-size statue of his beloved *Pithecanthropus*. He invented a special 'stereo-orthoscopic' camera, capable of taking photographs of fossils from all sides without distortion. But all his endeavours seemed to be doomed, which he found increasingly exasperating. The escalation of the controversy, the rejection of his views and the continuing sniping about the circumstances of the recovery and his status as an amateur were hard to bear, and personally hurtful for him. Instead of finding recognition for his incredible success, he found himself in the middle of a controversy that was not of his creation, and that overshadowed his achievement completely. As a result he became increasingly reclusive, and finally he — who had gone to extreme lengths to promote understanding of his finds, and carried them around with him in a suitcase for years to show to any researcher who might be interested — refused to receive any further scholars. In fact he is reputed to have hidden the fossils under the floorboards of his house. So deeply was he hurt that for thirty years he turned his back on the scholars. Opponents of evolution rejoiced, saying that this was attributable to his remorse, for the cardinal sin of having aided and abetted such a sacrilegious teaching with his *Pithecanthropus*.

Aged 74 he relented at last and invited several prominent scholars to see him. A new generation had taken over in the discipline, whose methods and ideas were somewhat more sophisticated, and who were no longer concerned about the issues that had dominated Dubois' thinking. The Javan fossils were now attributed to *Homo erectus*, numerous specimens of whom had in the meantime been discovered on a hill at Zhoukoudian near Beijing. At 80, Dubois was visited by Franz Weidenreich (1873–1948), who was to become the most significant palaeoanthropologist in the discipline's history (and whose most important teaching has to this day been completely misunderstood by most Anglophone specialists, in what could be described as the discipline's major misunderstanding). This was shortly before his formulation of the multiregional hypothesis of hominin origins (Weidenreich 1947), which has now been disparaged for decades. The issue of the 'missing link' was no longer so pressing, after all, that link had been discovered in England, at Piltdown (which we will examine in the next chapter). In fact even a kind of 'missing link' that *was not a fake* had been discovered by that time, in South Africa. The only problem with that was that nobody took any notice of it.

## The discovery of *Australopithecus*

The Piltdown affair had a considerable effect on the discipline, particularly in the rejection of evidence that would tend to contradict Piltdown. Even evidence that might detract from the importance of 'Piltdown Man' was unwelcome for several decades. So when in 1924 a young, Australian-born anatomist in South Africa reported finding skeletal remains of a creature that seemed about half-way between ape and human, his report was greeted with scorn and contempt. Having in the previous decade discovered that humans had evolved in England, European and especially British archaeologists and physical anthropologists were in no mood to seriously consider such a competing counter claim 'from the colonies'. The infant specimen from Taung, in Bophuthatswana, consisted of a brain cast and the facial bones of a creature Raymond Arthur Dart (1893–1988) named *Australopithecus africanus*, the southern ape of Africa (Figure 21). He had found them in two crates of fossil bones from a limestone quarry. The Taung child (Dart 1924) had distinctive human-like features, and yet the experts of the time relegated it to the status of a new fossil ape. After all, Piltdown made it perfectly clear that the brain evolved before the rest of the skull did, and only after it matched that of modern humans did the remaining bone structure develop in this direction. On that basis *Australopithecus*, despite its rather human-like dentition, still had to be an ape.

*Figure 21. Raymond Arthur Dart.*

During the 1920s, tonnes of *Australopithecus* bones were quarried at another South African site, Makapansgat, and burnt to make lime powder. A local naturalist named Eitzman tried desperately to interest a palaeontologist in the study of the incredible wealth of fossils in this former cave's fill material. He made limited observations the best he could, and reported seeing an apparently complete skeleton of *Australopithecus* before it was thrown into a kiln. All his endeavours to interest anyone in this site were in vain, and when he showed the Makapansgat jasper cobble to Dart, who was himself disappointed by the reception *Australopithecus* had received, Dart expressed no interest (Figure 22). This stone, from an australopithecine-bearing deposit almost 3 million years old (McFadden et al. 1979), is now regarded as the earliest evidence ever recovered that suggests the emerging capacity (pareidolia) to perceive figurative properties in a natural form (Eitzman 1958; Dart 1974; Bednarik 1998). This important piece of evidence was not properly studied and analysed for 72 years (i.e. until I did so in 1997). Indeed, many aspects of the

*Figure 22. The Makapansgat jasperite cobble, a late Pliocene manuport.*

discovery of *Australopithecus* indicate an incredible scientific apathy.

A medical doctor, Robert Broom, who had been the only one supporting Dart's view, found further specimens of this very early, human-like creature in the 1930s and 1940s, and towards the middle of the century it was becoming glaringly apparent that Piltdown was a fake. It was also becoming clear that its acceptance by the experts had retarded the study of human origins for four decades, and that there was more than one species of *Australopithecus*. Today we distinguish several of them, and we think that the australopithecines persisted for millions of years in Africa. Fortunately for Dart, he was a young man when he discovered them, and he survived the time of the eventual acceptance of his find by many years. Dart did travel to London in 1930 to try and win support for *Australopithecus*, but without success. He gave up the study of fossils and focused instead on his work at the Witwatersrand anatomy department. At least he had not been victimised, his reputation was not torn to shreds; he had merely been ignored. As Boucher de Perthes had observed: "They did not discuss my facts, they did not even take the trouble to deny them. They disregarded them."

## Modern heretics in archaeology

Having thus examined key examples of the treatment mainstream archaeology has meted out to heretics, two observations may be made. First, certain uniformities are beginning to emerge in the way the discipline deals with troublesome heretics. Secondly, one might ask: is there a decrease in the severity of their treatment as we approach more recent, and supposedly more enlightened times?

We have quite a number of such cases of modern times. One of the most glaring examples in Pleistocene archaeology is that of Alexander Marshack (1918–2004), an American author who when aged around fifty was assigned to write about Palaeolithic art in Europe (Figure 23). His archaeological knowledge at the time was negligible, but as a very original thinker he soon noticed that this body of evidence was studied by antiquated methods and in surprisingly unscientific ways. For instance, he developed a strong interest in the work processes involved in engraved markings on decorated portable plaques, after realising that they could tell us a great deal about the circumstances of these productions. Until the 1970s, archaeologists had made no consistent use of microscopy in the study of archaeological work traces. The only exceptions were in Russia where S. A. Semenov (1964) had developed the microscopic study of used stone tool edges in determining what Upper Palaeolithic artefacts had been used for. Marshack introduced this kind of technology to the study of portable art objects and called it 'internal analysis'. For the next thirty years he produced a substantial series of exceptionally scholarly publications, explaining in some detail how he deduced certain circumstances of the manufacture of early markings from optical microscopy (Marshack 1972, 1975, 1977, 1984, 1985, 1989). While it is true that all scientific evidence is presented for the purpose of review, and indeed, for falsification, it is equally true that Marshack's work was subjected to far more critical attention than that of comparable work by a professional archaeologist. Yet no significant part of his hypotheses was ever falsified by an archaeologist, and particularly Francesco d'Errico, who embarked on a sustained campaign to test Marshack's findings, ended up agreeing with them substantially. Still, acceptance of these findings by the mainstream discipline was always grudging, and Marshack remained an outsider and was not accorded formal recognition. This was not because his scholarship was in question, rather the reverse. His mistake was his scholarly and restrained tone of discourse, and that he sought to address the savants rather than the public. Nothing he could do and nothing he could present would ever change their attitudes to him, which were not governed by the merits of his work, but entirely by his status as an amateur. His work and his findings are revolutionary for archaeology, and if they had come from an establishment archaeologist they would be valued and well received. Coming from a member of the archaeologically most despised group, archaeological amateurs, they had to be greeted with disdain, and ways had to be found to disprove them. After all, professional archaeologists were not in the business of accepting corrections by mere amateurs.

*Figure 23. Alexander Marshack.*

In the mid-1990s we experienced 'Altamira in reverse'. This time around, Pleistocene archaeology was the champion of the notion that a body of rock art in Iberia, found at a series of sites of schist exposures, are the open-air version of Pleistocene cave art (Bahn 1995). This issue came to a head with the announcement, at the end of 1994, of supposedly Palaeolithic petroglyphs in the lower Côa valley of northern Portugal. Along the present river course, just a few metres above the water, there is a series of impact petroglyphs and engravings on perfectly flat, slate-like vertical panels. They are mostly very well preserved, occurring generally in the close vicinity of the ruins of water mills dating from recent centuries. Among clearly sub-modern motifs occur animal pictures, which appear to depict horses, Spanish fighting bulls and a few goats or ibex. They are generally no more patinated and weathered than the inscribed dates at the same sites, in fact the 18th century dates are consistently more weathered than the semi-naturalistic animal images (see Chapter 5 for details). When I, as an amateur archaeologist, opposed the unanimous archaeological mantra of the Palaeolithic age of this rock art in 1995, I had no anticipation of the consequences. Like de Sautuola, I was attacked with a ferocity that almost defies description. Dozens, even hundreds, of Pleistocene archaeologists ridiculed, defamed and denounced me — the upstart who had dared not only to defy their authority, but also questioned the discipline's competence by writing about its epistemology: the processes by which it acquired its perceived knowledge.

Rather than silencing me, this treatment only prompted me to turn my attention to other blunders of the discipline. Next, I pointed out that the shamans of archaeology had for almost forty years remained ignorant of the fact that a Dutch non-archaeologist, Theodor Verhoeven, had in the late 1950s demonstrated the presence of hominins together with a stegodon-dominated fauna on the Indonesian island of Flores (Verhoeven 1958). The age of this human presence on an island that had never been connected to any other landmass was estimated at between 700,000 and 800,000 years (as shown by Ralph von Koenigswald, see Figure 26), and it had gone unnoticed because the finds had not been published in English until the 1990s (Bednarik 1997). Therefore it had not been appreciated that the humans concerned, *Homo erectus*, must have had the ability to cross the sea in adequate numbers to render the founding of new island populations possible. This species had not only colonised Flores, and necessarily also Lombok and Sumbawa on the way there, but also, as I demonstrated in 1998, Timor. To determine the minimum technological capability for these, and many other Pleistocene sea crossings demonstrated, I built eight very primitive rafts with stone tools and attempted to sail them (Figure 24). The archaeologists still sought to oppose my evidence, but no longer with quite the same zeal as before (Bednarik 1999, 2001, 2003; Bednarik and Kuckenburg 1999). They were beginning to realise that I outperformed them in published debates, that they had been less than knowledgeable, and that I would expose their shortcomings if prompted. Indeed, I continued tabling example after example of ignorance and failure on the part of Pleistocene archaeologists. The more models and claims I investigated, the longer the list of my refutations and corrections grew. It began to emerge how often the paradigms of this discipline are based on the ignorance of their promoters, and on a surprisingly sloppy epistemology.

The most recent major example of my falsifications concerns the persistent claims that all extant people derive from a small population in sub-Saharan Africa, whose descendents displaced or exterminated all other humans

*Figure 24. The stone-tool built bamboo raft* Lombok *off the coast of Flores, Indonesia, in 2008.*

by about 30,000 years ago and took over the world. This is known as the replacement model, or as the African Eve model. I demonstrated that it lacks any genetic, palaeoanthropological, cultural or even archaeological support, and that alternative models are very significantly more plausible and better supported by the hard evidence (see Chapter 5). I replaced the 'replacement model' with the 'domestication theory', and at the time of writing this book the response of the discipline to this new challenge has yet to be expressed.

Certainly, the many errors, blunders, controversies and shortcomings I have now exposed in Pleistocene archaeology render an examination of this discipline essential. How did it develop, how did it acquire its practices, its dogma? Indeed, if we consider the history of the discipline, outlined in the historical examples I have given above, it is justified to ask why such an investigation of its epistemology has not long been attempted. If we are to secure a better understanding of the distant human past, of our evolution, our development as a species, is it not necessary to confront these issues squarely and honestly? What we have seen so far is largely a caricature of Pleistocene archaeology (see following chapters), and the practices of knowledge acquisition, processing and presentation in this field need to be investigated thoroughly.

Anyone doubting the need for this, or my proposition that nothing has improved in the way the discipline operates, might like to consider just one very recent example. Prompted by me, Indonesian and Australian archaeologists discovered a tiny human creature in 2004. Excavated in the cave Liang Bua in western Flores, the remains were named *Homo floresiensis*, also known as the 'Hobbit'. Only about a metre tall, the creature immediately became the object of a major academic controversy, and interpretations of the find now range from gibbon to microcephalic modern human, and include several versions between these two extremes, such as *Australopithecus*, habiline and erectoid. What this extreme spectrum of opinions shows most clearly is that experts lack the ability of identifying human remains reliably at the species level. Any intelligent person can see that the bones are of a human-like primate, so if the combined experts of the world cannot find agreement on where in this wide spectrum they rightfully belong, how can one take seriously their claims, for instance, that modern people and Neanderthals could not breed? The obvious conclusion is that they are always just guessing. But what is particularly relevant is the all too familiar pattern: the accusations of theft and misconduct, the rubbishing of opposing views and individuals, the clamouring for attention, the ridicule, the unscholarly behaviour of the protagonists. Indeed, the similarities with the historical precedents are the most incredible aspect of this episode, demonstrating that the discipline has in the last 180 years learnt nothing from its own mistakes. If it is to improve at all, its epistemology must be put under the microscope.

## REFERENCES


Bahn, P. 1995. Cave art without the caves. *Antiquity* 69: 231–237

Bednarik, R. G. 1997. The initial peopling of Wallacea and Sahul. *Anthropos* 92: 355–367.

Bednarik, R. G. 1999. Maritime navigation in the Lower and Middle Palaeolithic. *Comptes Rendus de l'Académie des Sciences Paris* 328: 559–563.

Bednarik, R. G. 1998. The 'australopithecine' cobble from Makapansgat, South Africa. *South African Archaeological Bulletin* 53: 4–8.

Bednarik, R. G. 2001. Replicating the first known sea travel by humans: the Lower Pleistocene crossing of Lombok Strait. *Human Evolution* 16(3–4): 229–242.

Bednarik, R. G. 2003. Seafaring in the Pleistocene. *Cambridge Archaeological Journal* 13(1): 41–66.

Bednarik, R. G. and M. Kuckenburg 1999. *Nale Tasih: Eine Floßfahrt in die Steinzeit.* Thorbecke, Stuttgart.

Boucher de Perthes, J. 1846. *Antiquités celtiques et antédiluviennes*, Paris.

Cartailhac, E. 1902. Les cavernes ornées de dessins. La grotte d'Altamira, Espagne. 'Mea culpa d'un sceptique'. *L'Anthropologie* 13: 348–354.

Dart R. A. 1925. *Australopithecus africanus*: the man-ape of South Africa. *Nature* 115: 195–199.

Dart, R. A. 1974. The waterworn australopithecine pebble of many faces from Makapansgat. *South African Journal of Science* 70: 167–169.

Darwin, C. 1859. *On the origin of species by means of natural selection, or the preservation of favoured races in the struggle for life.* John Murray, London.

Eitzman, W. I. 1958. Reminiscences of Makapansgat Limeworks and its bone-breccial layers. *South African Journal of Science* 54: 177–182.

Fuhlrott, C. J. 1859. Menschliche Überreste aus einer Felsengrotte des Düsselthals. Ein Beitrag zur Frage über die Existenz fossiler Menschen *Verhandlungen des Naturhistorischen Vereins Preussen und Rheinland Westphalen* 16: 131–153.

Lyell, C. 1863. *The geological evidence of the antiquity of man.* John Murray, London.

McFadden, P. L., A. Brock and T .C. Partridge 1979. Palaeomagnetism and the age of the Makapansgat hominid site. *Earth Planetary Science Letters* 44: 373–382.

Marshack, A. 1972. *The roots of civilization.* New York: McGraw-Hill/London: Weidenfeld and Nicolson.

Marshack, A. 1975. Exploring the mind of Ice Age man. *National Geographic* 147: 62–89.

Marshack, A. 1977. The meander as a system: the analysis and recognition of iconographic units in Upper Palaeolithic compositions. In P. J. Ucko (ed.), *Form in indigenous art. Schematisation in the art of Aboriginal Australia and prehistoric Europe*, pp. 286–317. Australian Institute of Aboriginal Studies, Canberra.

Marshack, A. 1984. The ecology and brain of two-handed bipedalism: an analytic, cognitive and evolutionary assessment. In H. L. Rotitblat, T. G. Bever and H. S. Terrace (eds), *Animal cognition*, pp. 491–511. Lawrence Erlbaum, Hillsdale, N.J.

Marshack, A. 1985. Theoretical concepts that lead to new analytic methods, modes of enquiry and classes of data. *Rock Art Research* 2: 95–111.

Marshack, A. 1989. Methodology in the analysis and interpretation of Upper Palaeolithic image: theory versus contextual analysis. *Rock Art Research* 6: 17–38.

Prestwich, J. 1859. On the occurrence of flint-implements, associated with the remains of extinct mammalia, on undisturbed beds of a late geological period. *Proceedings of the Royal Society of London* 10: 50–59.

Sautuola, M. Sanz de 1880. *Breves appuntes sobre algunos obietos prehistóricos de la Provincia de Santander.* Martínez, Santander.

Semenov, S. A. 1964. *Prehistoric technology* (transl. M. W. Thompson). Londres, Cory, Adams and Mackay, London.

Schwalbe, G. 1901. Der Neanderthalschädel. *Bonner Jahrbücher* 206: 1–72.

Verhoeven, T. 1958. Pleistozäne Funde in Flores. *Anthropos* 53: 264–265.

Weidenreich, F. 1946. *Apes, giants, and man.* University of Chicago Press, Chicago.

# 5

# MISTAKES IN ARCHAEOLOGY

In Chapter 4 we have briefly touched upon the question of mistakes that have been made in archaeology historically — but we have considered these only in the context of specific issues at hand, such as the treatment of dissenters who were eventually shown to have been right. Here we will consider more carefully the generics of this problem: are there definable patterns in the epistemology of mistakes? Can we identify common or systematic factors, or do blunders in archaeology occur in random patterns? Since the purpose of this exercise is not to discredit individuals, but rather to strengthen through constructive critique, I will provide no names of those who became victims of their own overconfidence (except where the actors have long died) and not reference recent archaeological bungles. I will also focus on case histories in which not a small group or an individual failed, but where most or even all of the discipline shares responsibility for the mistakes. The objective here is to examine what features or currents of archaeology are conducive to blunders of wide-ranging effects, and particularly to explore what specific aspects tend to facilitate the development and entrenchment of the trends and fads archaeology appears to be subject to.

## Piltdown man

One of the most celebrated frauds in the history of archaeology (although perhaps not its most consequential) concerns the find of nine fragments of a brain case and a right lower mandible with two molars, found in a gravel pit in Sussex and unveiled to the 'scientific' world in December 1912. The Piltdown discovery consisted of an ape-like jaw with a seemingly modern human cranium (Figure 25), and since archaeology was actively looking for a 'missing evolutionary link' at the time of the early 20th century, the find was readily

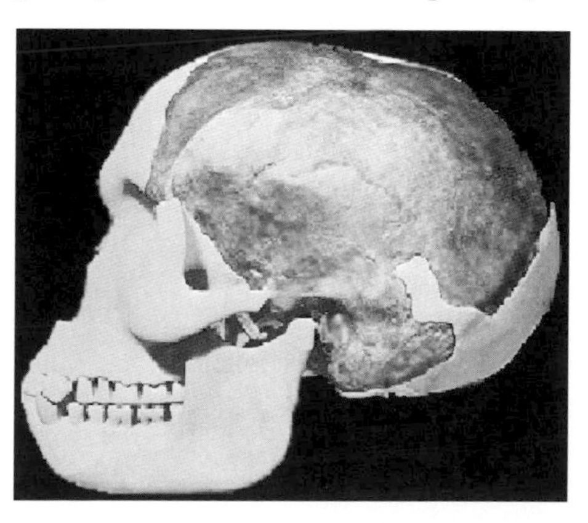

*Figure 25. The Piltdown hoax, comprising a modern human skull and an ape mandible.*

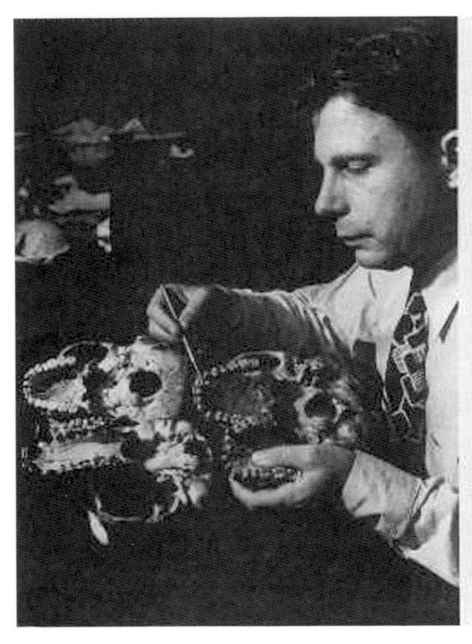

*Figure 26. Gustav Heinrich Ralph von Koenigswald (left) and Franz Weidenreich.*

accepted. A few scholars were sceptical from the start, and during the 1940s, 'Piltdown man' became increasingly anomalous. Thanks to the detective work primarily of Kenneth Oakley, a non-archaeologist, its fraudulent nature was ascertained in 1953 (Weiner et al. 1953). However, it is often overlooked that Franz Weidenreich (Figure 26) examined the remains in the 1920s and already then correctly determined that they comprised a human cranium and an orang-utan jaw with filed-down teeth. In the meantime, this false evidence had a great impact on the study of human origins. For instance when Raymond Dart reported his Taung australopithecine fossil in 1925 it was greeted with contempt in Britain, as noted in Chapter 4.

The culprit(s) of the Piltdown fraud was (were) never established, although Charles Dawson (the co-discoverer) was long suspected. After all, he was an amateur archaeologist, which already renders him suspect in the eyes of 'real' (paid) archaeologists. There are probably as many theories about who perpetrated the fraud and for what purpose as there have been commentators. Among the principal suspects, apart from Dawson and Arthur Smith Woodward (the other co-discoverer), were Sir Arthur Keith, who was conservator of the museum at the Royal College of Surgeons and had spirited debates with Smith Woodward over how the fragments should be assembled; Australian Sir Grafton Elliot Smith, who was favoured by Ronald Millar in his book, *The Piltdown men*; Professor William Sollas, a geologist from Oxford (accused by J. A. Douglas, Smith Woodward's successor at Oxford); Teilhard de Chardin, because he was a cleric, and because of certain comments he made; and even Sir Arthur Conan Doyle (the creator of Sherlock Holmes), who happened to live nearby and had been to the site. In Australia, Ian Langham favoured Elliot Smith as the likely culprit, perhaps together with palaeontologist Dr Smith Woodward as his accomplice. Others named as suspects include Lewis Abbott, Frank O. Barlow, William Butterfield and V. Hargreaves. An interesting variation to a common theme was that Dawson, a lawyer, initiated the fraud, which was spotted by Hinton who sought to protect his superior, Smith Woodward, from ridicule, and tried to expose Dawson by subsequently planting more outrageous objects, such as a bone shaped like a cricket bat. The clumsiness of the fraud seems to be intentional, and appears to exclude Dawson as the perpetrator: if the fraud had been intended to remain undetected, it would have been executed much better. Rather than a fraud this appears to be a hoax, with Smith Woodward the intended victim (Figure 27).

In the 1990s British researchers Brian Gardiner and Andrew Currant discovered fossil elephant and hippopotamus teeth and various bones in a trunk stored in the Natural History Museum (Gardiner 2003). The remains had been stained in the same way as the Piltdown material, and the trunk was marked with the initials of Martin A. C. Hinton, who had been volunteer curator of zoology at the time. He had already at the age of sixteen published a paper on how to impregnate bones with iron and manganese oxides to make them appear ancient. The relative proportions of iron and manganese were the same in the Piltdown finds and the objects found by Currant in the trunk, and both also contained chromium. The staining, it was argued, represents the evidence of a practice run, and it is claimed that Hinton wanted to dupe Smith Woodward because he bore a grudge against his superior who worked in the same institute.

*Figure 27. Portrait painted in 1915 of the examination of the Piltdown find. The discoverers, Dowson (left) and Smith Woodward are standing at the right.*

However, a variation of this explanation involves Pierre Teilhard de Chardin, who within weeks (5 January 1913) of the report of the find published a paper beginning with the statement '[t]here was a time when the study of prehistory deserved to be suspected, or the subject of jokes' (Teilhard de Chardin 1913). He noted that the mandibular condyle had been broken, which masked the bone's origin, and he subsequently (June 1913) 'discovered' an artificially patinated orang-utan canine at the Pitdown site, which apparently he had *earlier* shown Marcellin Boule (Clermont 1992). This and other information implies that Teilhard de Chardin, although perhaps not the initiator of the hoax, was involved and tried to expose it, albeit also unsuccessfully (Thackeray 1992).

If Hinton were indeed the originator, as appears rather likely on the basis of the available information (although there are also counterarguments), the fake evidence would not have been intended to fool anyone but Smith

Woodward himself. But it seems the hoax went terribly wrong: Hinton had not reckoned with the gullibility of scholars, who ignored even the most ridiculous additional material planted (and the cricket bat would surely imply a sense of humour on the culprit's part) and by the time he realised that his practical joke was accepted as real evidence by the discipline, instead of being exposed by it, he had no choice but to remain mute. So instead of damaging Smith Woodward's reputation and exposing his ignorance he ended up enhancing his status. One can only assume that it taught him a lesson: never underestimate the gullibility of 'scholars'. And if Teilhard de Chardin was also involved, as appears probable, he found himself in the same position of having to keep silent about a joke that badly backfired.

Seen in a historical perspective we can be grateful to Hinton (if indeed he was the perpetrator, which is not established), who has shown us that a healthy disrespect for the pomp and ignorance of academia is most requisite (Figure 28). But whatever the true answer to the Piltdown riddle is, most commentators would agree on one point: with a very few exceptions, all those hoodwinked by the hoax were genuinely duped; they really did believe in the find's authenticity. The situation was rather different in the next case.

## The trouble with Glozel

One of the most controversial archaeological sites is Glozel, in a small farming community near Vichy, central France. Discovered by a teenager, Émile Fradin, and his grandfather while ploughing a field called Duranthon on 1 March 1924, it comprises tombs or underground chambers, and has yielded many thousands of finds. These are mainly inscribed ceramic tablets, engraved stones, urns with faces, flint axes, apparent idols, fine bone sculptures, glass (including high-potassium glass typical of the Middle Ages) and skeletal remains. Systematic excavations were initiated by an amateur archaeologist, Antonin Morlet, in the following year. He published a report later in 1925, naming the young discoverer as his co-author and defining the site as 'Neolithic'. It was widely rejected by French archaeologists, no doubt because of the author's amateur status, but both Salomon Reinach, curator of the National Museum, and the famous abbé Henri Breuil

*Figure 28. Hinton (left) and Dowson at the Piltdown site.*

authenticated the finds. Breuil, after publishing two papers in 1926 declaring his support, then changed his mind and in October 1927, claimed that most of the finds were fakes, the exception being the ceramics. This occurred after some very curious developments. A close colleague of Breuil, André Vayson de Pradenne, approached Fradin under a false name and proposed to buy the collection. Fradin angered him by refusing and the archaeologist promised to "destroy the site". Indeed, he published a paper accusing Fradin of having salted the trench, after he obtained permission from Morlet to dig there (under his true name this time). Breuil, probably troubled by the fact that the reindeer, depicted on several items, had disappeared from France with the end of the Pleistocene, had identified an apparent reindeer engraving on a pebble as a 'cervid' (Figure 29). Morlet secured the judgment of the zoological director of a Norwegian museum that the engraving was of a reindeer, thus directly contradicting the 'Pope of Prehistory', which is when Breuil began rejecting Glozel.

The curator of th Louvre, René Dussaud, also accused the young Fradin of fraud, who responded by filing for defamation in early 1927 (having the pro bono support of a prominent attorney who was intrigued by the spectacular case of a 'peasant boy against the Louvre curator'). The Glozel inscriptions disproved Dussard's life work, which was based on the assumption that our alphabet is derived from the Phoenician. A commission excavating at the site late that year for three days pronounced all finds as fakes, with the exception of a few flints. By that time the issue had already descended into a farce. During this investigation, on 8 November 1927, Morlet observed three commission members slip under the site's barbed wire fence early in the morning, and then one of them, Dorothy Garrod, made a hole with her hand in the excavated section. Morlet immediately confronted her, she first denied

*Figure 29. Engraving of what appears to be a reindeer, with Glozel writing. This may possibly be a fake planted by archaeologists.*

the accusation, then admitted her action as there were several other witnesses (including an attorney and a science journalist, both of whom confirmed in writing having observed the incident). It is to be noted that Garrod, a distinguished archaeologist, had studied under Breuil, and like most members of this 'commission' was convinced the site was a fraud. Her indiscretion would not have come to light had the subsequent heated discussion on site not been photographed (Figure 30). The principal author of the commission's report, Count Bégouën, even had to admit having falsified part of it: one of its members, Professor Mendes Corréa, had supported the authenticity of Glozel, so Bégouën made up his commentary.

Next, the president of the French Prehistoric Society, Felix Regnault, filed a police complaint of fraud, leading to the confiscation of three cases of artefacts. During the police raid of his home, Fradin was beaten when he objected to the removal of his little brother's schoolbooks. Further excavations by a group of relatively neutral archaeologists took place in 1928, and they confirmed that the site was authentic. Nevertheless, in 1929 Fradin was indicted for fraud, but the ruling was quashed on appeal two years later. In March 1932, by which time the young farmer was 25, Dussard was found guilty of defamation and had to pay all court costs.

Much later scientific work has detected no evidence of fraud. Moreover, several sites with similar material are now known, including Chez Guerrier and Puyravel in the same region, and even as far away as in Portugal or Romania. The ages of bones from Glozel range from the 5th to the 20th century, while the ceramics fall into three groups, from 300 BCE to recent times. The flints are from all accounts 'Neolithic'. The most controversial components of the

*Figure 30. Morlet accusingly confronts Garrod, who hides behind several men and looks away guiltily, Glozel site, November 1927.*

substantial collection from Glozel are the approximately one hundred ceramic tablets bearing an unknown script. It has not been conclusively deciphered but appears to be of a Gaul dialect, dating apparently from the final Iron Age to the Roman period. It may derive from the Etruscan alphabet and vaguely resembles Phoenician script.

The inscriptions may well be of considerable value to the understanding of Gallo-Roman cultural connections, and their investigation seems to be more comprehensive than the archaeology of the site. Thermoluminescence analysis of many of the tablets places the majority of them between 300 BCE and 100 CE. It has been suggested that the characters resemble alphabets of the late Iron Age and the Gallo-Roman period, particularly the Lepontic script of northern Italy. The Lepontians spoke a Celtic language related to Gaulish. Gallic immigrants, who according to the Roman historian Livy migrated about 400 BCE to the Po basin, are thought to have adapted the Lepontic script, itself derived from a northern Etruscan alphabet. It is thought that they developed a Gaulish-Cisalpine alphabet, which in turn was imported into Transalpine Gaul. This is thought to be the source of the Glozel script (Figure 31).

The most likely interpretation of the substantial corpus of archaeological material from Glozel is therefore that it contained material of various phases, beginning with 'Neolithic' flints, but mostly dating from the last two millennia. The bulk of the finds, including the tombs, may be of medieval age. The issue remains unresolved because of the highly controversial status of the assemblage, which has prompted a lack of published data until recent years. It is quite possible that the material from Glozel includes fakes, because many of the players had to lose a great deal of prestige had its authenticity been proved. And many of them had the opportunity to salt the site — in fact one of them, Garrod, was caught attempting to do just that. She confessed many years later that her interference was to preserve "the honor of the discipline — allowing Glozel to be recognised formally would have damaged too many careers and reputations".

*Figure 31. One of the approximately one hundred ceramic tablets covered with writing, Glozel, considered authentic.*

More interesting than the character failings of many of the archaeologists involved are the legal and practical aspects of this affair. First of all, it should have always been obvious that it is absurd to accuse a teenager of rural background of perpetrating what would have been an extremely sophisticated scientific fraud. How could the young Fradin possibly have concocted such an elaborate scam, constructed underground chambers and salted the site with thousands of finds? How could he have invented an appropriate script, made the authentic tablets and the flints? Or erected the subsequently discovered nearby megalithic structures? The proposition seems absurd, yet the archaeological establishment sought to resolve the issue by legal sanction, intimidation and surreptitious means. Surely Fradin was not the one to be blamed for the inadequate excavations of the site that apparently lumped together materials of widely different antiquities. Surely the fault here lies with the archaeological establishment, with its confrontational approach and its pompous self-importance. An innocent young man who had accidentally stumbled upon the site had to fight for eight years to clear his name, while people of considerable authority used their weight trying to crush and destroy him. Similarly, Morlet, who continued his efforts to secure scientific authentication for Glozel until his death in 1965, died without having achieved it. The inability to untangle the mysteries of this site in several decades does not inspire great confidence in the many archaeologists who have contributed to this controversy.

Since it is absurd to attribute the alleged fake to Fradin, some commentators have even suggested that perhaps the site was salted well before its discovery, but this is again highly unlikely. If it had not been for the foot of Fradin's cow that broke into the first tomb, the Glozel site may still remain undiscovered. Moreover, it is hardly realistic that the tombs could have been constructed by unknown hoaxers, in an annually ploughed field. Retrospectively it is difficult to see how so many leading archaeologists could have erred so profoundly. There was never any indication of fraud, except that which archaeologists themselves probably contrived. The whole story defeats every purpose of archaeology, and seems to illustrate what happens when a discipline gives way to self-interest. Much the same occurred in the next example.

## The hot waters of the Côa

Now I would like to turn my attention to a much more recent and more serious controversy. A long-standing practice in Portugal of archaeological complicity in the destruction of rock art sites in that country was successfully challenged by Mila Simões de Abreu, who in December 1994 took on the archaeological establishment of Portugal. The destruction of cultural heritage is a worldwide phenomenon, but until that time it had never been challenged in any really consequential fashion. Archaeologists had long facilitated and often actually conducted the destruction of rock art and other immovable sites, instead of implementing the protective laws applicable around the world. Powerful developers and governments paid well for the clandestine eradication of such monuments that were in the way of development.

With the support of the local and international media, and that of rock art researchers around the world, through the International Federation of Rock Art Organisations (IFRAO), Abreu succeeded in stopping the construction of the Côa dam and the toppling of the Portuguese government in October 1995 (Figure 32).

But this was not, as one would have been entitled to assume, the end of the Côa controversy. A university professor specialising in the Palaeolithic period was appointed to head the restructuring of the discredited public archaeology of Portugal. Under his authority, an ardent effort to find evidence of Palaeolithic human presence in the Côa valley was commenced, particularly in the vicinity of the rock art panels so as to 'prove' their Palaeolithic antiquity. Sediments were churned up at over sixty sites in this ardent search and the rock art panels were systematically cleaned of their microflora. The wholesale removal of lichens with hard tools also removed fragile mineral accretions and rock flakes, and once it became known, Abreu and her supporters rightly voiced their criticism of these practices, labelling them as site vandalism. In the ensuing public brawl in the Portuguese mass media, the professor conceded that he had ordered the removal of the lichens. His admission was published in the national newspaper *O Independente* of 6 September 1996, under the title '*Broncôa. Zilhão admite limpeza das gravuras*' (*bronco* means stupid in Portuguese). He explained his actions as having been necessary due to "political expediency", and he offered to resign if a commission of four scholars, *nominated by himself*, would find the accusation of professional vandalism justified. It was well known that the professor was fervently committed to proving the Côa petroglyphs to be Palaeolithic, i.e. of the Solutrean. He vigorously and often abusively opposed the "fraudulent

*Figure 32. Part of the massive Côa dam construction site in June 1995, at the time it was abandoned.*

science" of archaeometry "imposters" and their "accomplices" who found the rock art to be quite young. That included me, I was one of three rock art dating scientists who were requested by the Portuguese government to test his age claims. When the discovery of the rock art was initially announced I had shared the view that it seemed to be of the Pleistocene, but after seeing detailed photographs and receiving a sample of the schist on which it occurred I counselled caution.

The need to remove lichens from the Côa sites is surprising, because lichen growth was not luxuriant, and there were no cases where lichens actually obscured petroglyphs. Lichen thalli in engraved grooves were mostly of microscopic size only. Sites where they were much better developed or had not been scrubbed before 1995 were never examined by the dating specialists, which is rather unfortunate. They would have offered numerous examples where large thalli were actually dissected by engraved or pounded lines, which shows that the petroglyphs must be younger than the lichens. Some of the greatest lichen concentrations had occurred at the site Barca, where two distinctive species formed discrete areas of growth. The engraved animal figures, where they cross lichen thalli, always postdate these zones, but some small thalli have developed in their grooves locally, postdating the petroglyphs (Figure 33). By sampling the two sets of lichens, those older and those younger than the petroglyphs, scientists could have acquired dating information bracketing the art with precision and reliability. Selective sampling of dead lichen tissue could have probably achieved dating of the art to within some tens of years.

Once the professor, whose views were shared by practically all the world's commenting archaeologists, had taken up his position as director of Portuguese archaeology, he would have discovered the lichen occurrences that would disproof his hypothesis of Palaeolithic age. Soon access to the Barca sites was limited to his own staff and associates, and numerous requests for access by other researchers were rejected. An American rock art conservation specialist, Jane Kolber, was forcibly ejected by a security guard from the neighbouring site Penascosa on 29 July 1996. Why all the secrecy, and why ban observers from excavations?

There can be no doubt that the rock art could have been dated through the lichens at five panels of the two Barca sites. Their physical and chronological relationship to the petroglyphs is obvious even to the layperson. The professor claimed that he had the lichens systematically removed because of some misguided site management policy, and because they might pose a threat to the rock art. He also claimed that he believed the rock art is well in excess of 20,000 years old. The Côa petroglyphs are exceptionally well preserved, so why did the removal of the lichens become so important after 20,000 years?

The professor's systematic scrubbing of the lichens with "wooden tools and river water", as he described it himself, has destroyed the possibility of determining the precise age of the petroglyphs via the related lichen thalli,

*Figure 33. Photograph published of a Barca petroglyph, Côa valley, Portugal, before the lichens were removed in 1996. The dissection of lichen thalli by engraved grooves is visible.*

and thus any testing of his stylistic dating of the art to the Solutrean may not be possible. These extraordinary developments in the Côa valley seem to be without precedent in the history of archaeology. Nevertheless, the mythology of the Palaeolithic age of the Côa petroglyphs continues to be supported by every commenting archaeologist in the world, and has been vigorously defended by many of them. The true age of the overwhelming majority of the valley's engravings and impact petroglyphs is in the order of one to four centuries, although a very few are older, and some may even date from the 'Neolithic'. In addition to numerous clearly recent images (of locomotives, bridges, fish, trees, clocks, crucifixion scenes and the like; see Figure 34) there are hundreds of inscriptions, many of them with dates. It is obvious even to the casual observer that these dates, mostly of the 18th to 20th centuries, are often much more weathered than the adjacent animal drawings of Spanish fighting bulls, horses (one shown wearing a bridle, see Figure 35) and ibexes the world's archaeologists contended to be of the Pleistocene. The professor also claimed falsely that ibex had not existed in the region since the Pleistocene, and that a colluvial sediment covering some petroglyphs dated from that period. Yet ibex still survive in the region, and in recent centuries were quite plentiful and a favourite target of hunters (Figure 36); and a colluvium is always younger than any of its components. In the particular case the sediment purported to be 26,000 years old was in fact less then 17 years old at the time! And a series of stone tools the professor had placed in the Upper Palaeolithic were 'Neolithic' microliths, which according to his own publications co-occurred

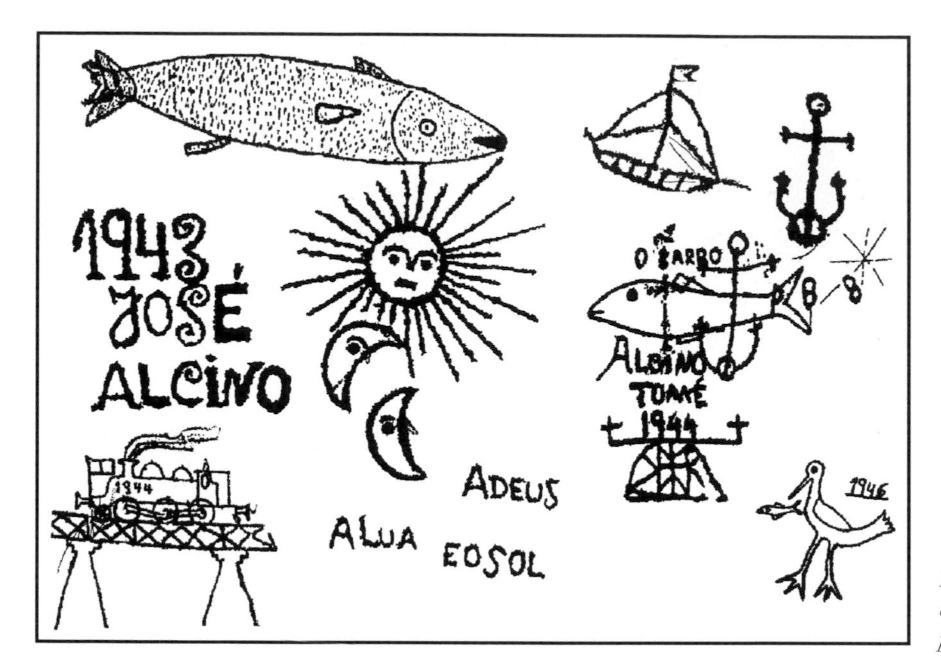


*Figure 34. Typical examples of Côa petroglyphs.*

with ceramics in practically all cases. Moreover, scientific dating of some of the petroglyphs by several archaeometrists, conducted as a blind test under government supervision (to exclude the possibility of collusion between the three analysts), yielded essentially identical results: the rock art was relatively recent. Finally, analysis of similar rock art at the site Siega Verde in the nearby Agueda valley has recently shown conclusively that the petroglyphs there, also claimed to be Palaeolithic, are all under 200 years old (Bednarik 2009a).

Nevertheless, the archaeologist in the centre of this issue was not the victim of some hoax; rather he became the victim of a fervent desire to prove the art's very great age, and in this quest he ignored all evidence conflicting with his view, no matter how persuasive it was. The lichen thalli dissected by the engraved grooves clearly rendered a Palaeolithic age impossible, and he probably realised that the lichens refuted his views. There are several lessons to be learnt from this controversy, the first relating to the possibility that some personal convictions may disqualify a person from exercising executive control over a site. In the present context of exploring the epistemological reasons for the Côa controversy, however, it is more important to review two other issues: why did a presumably experienced Pleistocene archaeologist convince himself that 18th century images are of the Gravettian, and why did an entire discipline believe him? Being able to answer these questions would

*Figure 35. Horse image at Fariseu, Côa valley, claimed to be about 26,000 years old; note the depiction of a bridle (horses were domesticated between 5000 and 6800 years ago).*

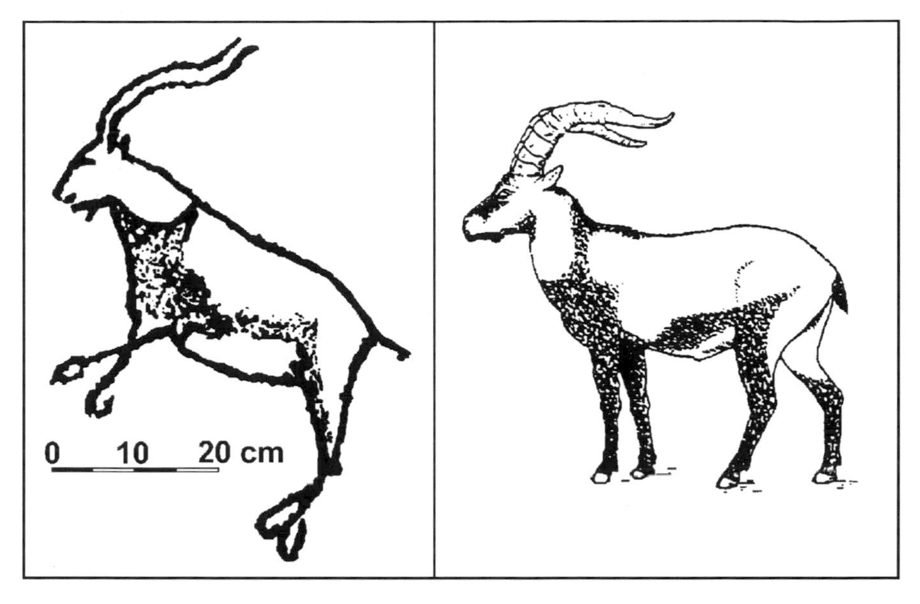


*Figure 36. Côa depiction of a caprid (left), falsely claimed to be extinct in the region since the Ice Ages, compared with* Capra ibex victoriae, *an extant subspecies found in the wider region.*

bring us a step closer to understanding what it is that renders archaeology, particularly Pleistocene archaeology, so accident prone.

It may be difficult to answer the first question, as we can never fathom the motivations of the individual, but the professor's objection to the "interference" of rock art scientists, called "interlopers" by him, seems to have initially led to his stance, and his inability to retract hasty pronouncements, having been made very publicly, perhaps prompted his intractability on the issue. This may therefore be a case similar to that of Cartailhac, which we visited in Chapter 4: relying on the often not refutable nature of archaeological propositions and on a personal hunch, a scholar finds himself in an academic cul-de-sac and, instead of backing out early, opts for seeking to 'preserve his reputation' by intransigence. This is then not necessarily attributable to the individuals' character, but rather to the academic conventions within which he operates. The Western system of academe encourages — even expects — the individual to defend a chosen position against all challenges. To admit that one was wrong is considered an academic defeat, and there is certainly a deficit of such concessions in archaeology.

The systematic acceptance of the Côa claims by the entire discipline is perhaps easier to explain, considering the available public commentary. Whenever a major scientific announcement is made in the mass media, its agencies solicit urgent comments from scholars that have in the past tried to catch the attention of the media. The researchers concerned then have the choice of admitting that they lacked the adequate information about the news item to comment, or they can quickly assemble a hasty response to enhance their public exposure. In this case the claims were accepted prematurely (before any details of the find had become available) by several attention-seeking students of Palaeolithic rock art. Publicly influential scholars determine dogma, and the discipline then amplifies these views. Whenever challenges arise, they are then strongly contested to preserve the credibility of those who have committed themselves prematurely, and most particularly so when they derive from 'outsiders': practitioners from other disciplines, and especially through the "impertinent transgressions" of non-archaeologists.

## The Jinmium blues

While we are on the subject of very old rock art: 'much older' petroglyphs have been reported in northern Australia, from the Jinmium site in the eastern Kimberley region, Western Australia. The considerable media hype around the world that followed the announcement in 1996 of this "oldest rock art in the world" could have easily led to a stalemate similar to the Portuguese example we just considered. Here, however, the controversy developed very differently, and this was purely the result of the particular mode of involvement of the press.

Jinmium is a small sandstone shelter whose wall carries a large number of cupules. These are hemispherical depressions hammered into the rock surface, a form of rock art that occurs globally and tends to be among the earliest manifestations of non-utilitarian or symbolic activities. Therefore the Jinmium rock art might have been a good contender for great antiquity. The sediment deposit in the shelter was excavated (Figure 37), thermoluminescence 'dates' were secured from it, and an exfoliated rock fragment with cupules was found at a level corresponding to TL dates of 58,000 to 75,000 years ago. Based on the TL data it was also claimed that the site's earliest human occupation occurred about 176,000 years ago — three times as long ago as the hitherto accepted first colonisation of Australia. Instead of presenting these extraordinary claims for peer review in the academic literature, as should properly have been done, they were announced in an exclusive newspaper report in *The Sydney Morning Herald* and the Melbourne *Age*, two broadsheets of the same corporate entity, on 21 September 1996.

This form of public announcement had several effects. First of all, other newspapers cried foul, saying that the research had been publicly funded and the announcement of its results should therefore have been made to all media, not just to two related newspapers (Rothwell 1996a, 1996b, 1996c). Secondly, the premature public announcement deprived the project of the safety net of open peer review. The 'scientific' article reporting the Jinmium claims was to appear two months later in the British journal *Antiquity*. So I suggested to the editor of that journal that he was entitled to cancel publication of the paper, because of the premature media hyperbole the material had been subjected to in the mass media. The editor responded by telling me that "he was proud to be a member of the research team" and would run the article. My subsequent suggestion that he publish with the paper also an opposing view by an Australian university professor, to provide balanced coverage for

*Figure 37. The cupules of the Jinmium shelter in the far north of Western Australia.*

what was becoming a very controversial theme, was similarly rejected — a mistake that eventually cost the editor his job. Some of the newspapers that had been slighted by the selective announcement began to test the claims by the Jinmium team, which rightly would have been the responsibility of the discipline, through refereeing the paper. So here we have the trial of a spectacular archaeological claim by the media, instead of the discipline, because the latter had not conducted such testing. Most of the media circus following the premature announcement of the Jinmium findings consisted of recitals of archaeological platitudes and countless errors of fact. For instance the most spectacular specific claim was that the Jinmium cupules are the oldest rock art in the world. The oldest *known* rock art in the world is in Auditorium Cave and Daraki-Chattan, India (Bednarik 1993; Bednarik et al. 2005), and is safely attributed to the Lower Palaeolithic (well over 170,000 years old; see Figure 38).

A team of archaeometrists eventually conducted scientific dating and reported its results in May 1998. The problem with the TL method is that it seeks to determine the time mineral grains were last exposed to light, which would correspond to the time they became concealed by further sediment deposition. However, where the sediment is derived from the quartz grains of decomposing sandstone fragments that had exfoliated from a rock exposure above, only the grains on the surface of each fragment could have been exposed to light. As these clasts disintegrate in the soil the grains in their interior are freed, but the TL results from them suggest extremely high ages. The way to overcome this problem is to determine the last exposure time of only those grains that were actually subjected to light. This is done by analysing each sample grain individually, and then ignoring those values that are clearly much too high. This method is called optically stimulated luminescence method (OSL) and it yielded in the case of the Jinmium site perfectly sensible results, matching carbon isotope results. The sediment was entirely of Holocene age, i.e. less than 10,500 years old, and the oldest human occupation was estimated to be in the order of 6000 years ago, rather than 176,000 years. The exfoliated rock art fell to the ground around 4000 years ago (Roberts et al. 1998). The entire issue was the result of misuse of a dating method for an application it was not intended for. It was a great embarrassment for Australian archaeology, and it could have been avoided easily. Fortunately it did not lead to a protracted controversy — at least partly because of positive press involvement.

## 'African Eve': a gene fetish

*Figure 38. Cupules of the Lower Palaeolithic of India, at Daraki-Chattan.*

A classical example of an archaeological blunder that held the discipline enthralled for well over twenty years, and that at its zenith held sway for almost all practitioners (although never quite managing to silence the last few dissenters), is the following highly improbable scenario. A new human species, miraculously evolving separate from any other populations and thus becoming unable to produce fertile offspring with other humans, arose in sub-Saharan Africa, and then expanded throughout the occupied world, replacing or exterminating all other humans in their wake. These 'Moderns', as they were called, could all trace their genetic lineage back to one single female, which the mass media facetiously dubbed the 'African Eve'. This strange tale is called the replacement theory in archaeology, and one of its key postulates is that the 'African Moderns' invaded Europe around 35,000 years ago and replaced (either exterminated or out-competed) the resident 'Neanderthal people'. Indeed, these African *Übermenschen*, our ancestors, are said to have wiped out all other people of the world. Importantly, there is no archaeological evidence in favour of this model and it is based purely on the views of some geneticists, rejected by other geneticists. Its foundations are fictional DNA base-pair substitution rates, unknown population sizes and incorrect assumptions about unique colonisation events (Bednarik 2008a, 2011); therefore it does not appear to have any demographic justification. Despite these significant shortcomings, this model has for over twenty years provided the dogma for the origins of 'anatomically modern humans', especially in the globally dominant Anglo-American version of archaeology (see Chapter 2 concerning the regional fragmentation of the discipline).

Before this genocidal model's weaknesses are considered, its obvious shortcoming is the way in which it addresses human modernity. It links it to specific skeletal differences (mostly concerning crania) and controversial genetics, when in fact it should be obvious that human modernity is primarily an issue of cognition, intelligence and culture. What renders hominins so different from all other primates, extant or extinct, are these factors, and not appearance or skeletal characteristics. So the model is initially couched in questionable premises, before we even consider its actual merits or problems. In a scientific sense, the question of 'anatomically modern humans' is an anthropocentric triviality; we do not search with the same enthusiasm for the origins of the anatomically modern fruit fly, because the first consideration of biology are taxonomic issues of separating species.

But when I began to examine the replacement theory more closely it became also evident that it was based on a previous model by a German palaeoanthropologist, G. Bräuer, called the Afro-European *sapiens* hypothesis. This, in turn was largely derived from a series of radiocarbon datings of human remains from Europe and the hypothesis of R. Protsch. In 2003 it was discovered that most of these datings were false. Several had been provided by Professor Reiner Protsch 'von Zieten' and were as bogus as that man's aristocratic title and second doctorate (he was heavily fined by a German court for the second imposture). Protsch had supplied the discipline with false ages of skeletal remains for about 30 years before it was discovered that he could not operate the analytical equipment, and that he had simply guessed all of his results (Schulz 2004). They turned out to be spectacularly wrong, for instance the Paderborn-Sande skull fragment, dated by him at about 27,400 years, turned out to be only 238 years old. Although this affair is just as damaging to the discipline as the Piltdown affair was, it remains almost unknown outside of Germany; it was essentially hushed up by the discipline to protect the reputations of those that had been hoodwinked. Apart from several publications by me, there is no comprehensive discussion of its effects in the academic literature.

Moreover, this development coincided with the announcement that several further human remains attributed to the crucial period from about 35,000 to 28,000 years ago — the latter date marking the time when the 'Neanderthals' were said to have finally disappeared — were also false (Figure 39). These fossils, too, were younger than the dogma of human evolution had demanded. In fact by 2006 it became obvious that there are no 'anatomically modern' human fossils of more than 27,700 available in Europe, i.e. from the entire first half of the Upper Palaeolithic period — when the replacement advocates had consistently claimed that this technological change was introduced in Europe by the 'modern' African invaders (Bednarik 2007). In contrast, human remains defined as 'Neanderthals' were being found from this period, and I proposed that even the rock art and portable art of a period from 40,000 to 32,000 years ago (called the Aurignacian) is either the work of Neanderthaloid people, or of their immediate descendents. European humans experienced a gradual change from very robust cranial and other skeletal features to 'gracile' features, which began roughly 50,000 years ago and is still continuing today. For each ten millennia we proceed from the present time into the past, Europeans were roughly 10% more robust than they were at the end of each such interval, and there is in fact no indication of any sudden change in their morphology, at any point in time: Europeans descend from the 'Neanderthals' (Figure 40). Similarly, there is no sudden appearance of a new technology, of totally new types of stone tools or any such indication of a material culture introduced from outside. All of the cultural traditions we distinguish in the first half of the European Upper Palaeolithic (c. 45,000 to 27,000 years BP), of which we distinguish more than fifteen, developed locally and had no precedents in Africa. Similarly, art-like products such as beads and engravings had been in use for hundreds of thousands of years, and were not introduced from Africa a mere 35,000 years ago. Finally, even the genetic support tendered has dissolved: today we know that Neanderthal alleles persist in modern Europeans, Asians and Papuans (Green et al. 2010). And the finds from Denisova Cave in Siberia (Reich et al. 2010; Krause et al 2010) also demonstrate that my contention for the past couple of decades, that people intermediate between the Robusts and the Graciles existed in the Late Pleistocene, is correct; this is only the last nail into Eve's coffin.

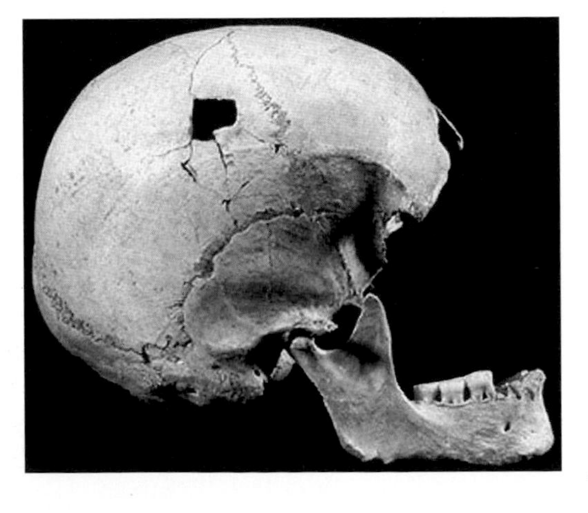

*Figure 39. Skull Stetten 2, Vogelherd Cave, Germany, widely claimed to be 32,000 years old. It is in fact only 4000 years old.*

Thus the entire replacement theory dissolved in a puff of smoke, and was even replaced with another, far more plausible model that has the support of the empirical evidence. This is the 'domestication hypothesis', which takes note of the fact that the gracilisation of robust humans is not a specifically European phenomenon, but it occurs in all four continents then occupied by humans, and at about the same time (Bednarik 2008b, 2011). This surely demands a systematic reason for such developments. The archaeology of the last 40,000 years recognises a rapid acceleration in cultural and technological development, and I have cited extensive evidence suggesting that cultural imperatives were becoming so strong that they gradually began to impact on mating patterns. Because gracilisation of humans begins in females, who still today are many millennia ahead of males in this development, a cultural

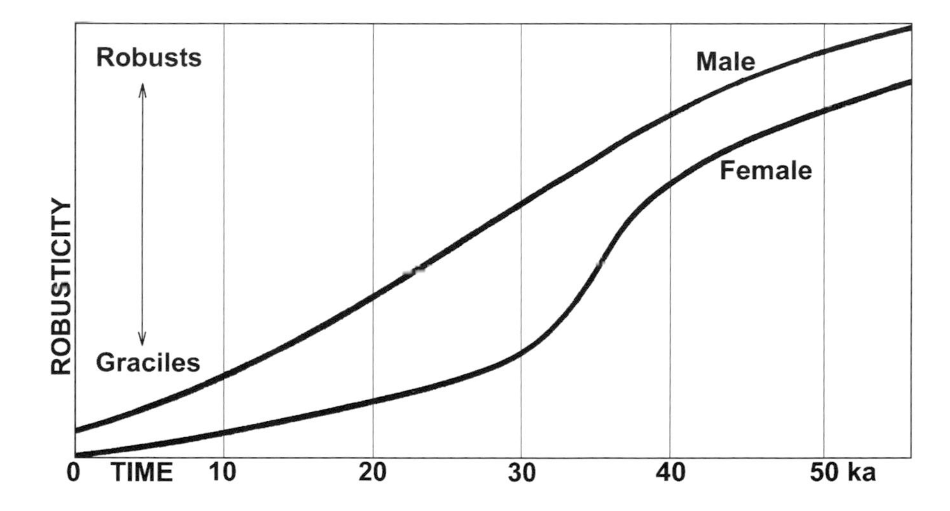


*Figure 40. Schematic depiction of male and female relative cranial gracility in Europe through time, showing that the decline in robusticity is gradual in males, but accelerated in females between 40,000 and 30,000 years ago.*

(conditioned) preference for neotenous females is suggested to have led to a gradual reduction in robust features. In other words, humans domesticated (i.e. influenced their own genetic makeup) themselves, albeit of course unintentionally. Biologically, the most outstanding features of today's humans are their state of encephalisation (the phenomenal development of the brain); their pronounced neoteny (adult humans resemble foetal chimps much more then these resemble adult chimps: we are essentially neotenous apes); and their unique susceptibility to thousands of deleterious genetic conditions. The unfavourable somatic effects of recent human 'evolution' involved not only a relatively sudden loss of robusticity, but also a loss of about 13% of brain size and up to one half of physical strength, and probably the loss of estrus. The gradual suspension of natural selection by domestication is reinforced by the even more dramatic appearance of neurodegenerative illnesses in modern humans, mental illnesses, and the toleration of literally thousands of Mendelian and other genetic disorders (Bednarik 2011).

Irrespective of this new hypothesis, the notion of an invasion of, first, the rest of Africa, then of Europe, Asia and Australia by a 'super-race' of genocidal people from sub-Saharan Africa is now fully superseded. It has no empirical foundation, and the idea that genes in modern populations indicate the mass movement of people in the past is absurd. Today's distribution patterns of genes are much easier to explain as the outcome of reticulate introgression and generational mating site distance. The important question that remains is to determine why the discipline has so enthusiastically embraced a rather cynical theory that seems to explain genocide and emphasises extreme competition among groups of humans. Actually, this was again much more pronounced in the Anglo-American school than elsewhere. Most notably the Chinese tradition of hominin research has steadfastly rejected the replacement theory all along, and several other schools were only lukewarm in their support. It would then seem that the British school is particularly prone to fads in Pleistocene archaeology. This is apparent from several developments, such as the old chestnut of whether there is any Pleistocene rock art in Britain. Perhaps there is, but the issue is firstly not particularly important, and secondly the documentation of such claims has, over an entire century, always been inadequate (Bednarik 2009b).

So why was the African Eve so enthusiastically welcomed, especially in Britain? The answer is to be found in the books of her most ardent supporters. They were unaware of much if not all the contradictory evidence, and they mistakenly conflated the issues of skeletal morphology with technological and other changes: according to them, assumed 'modern' anatomical appearance equals modern cognition and more sophisticated tools. But none of the developments in technology or art production coincide with the purported appearance of the 'Moderns' from Africa. For instance it has been known since 1979 that the 'Neanderthals' of a tradition called the Châtelperronian used art objects, so this was explained away by the absurd proposition that they must have scavenged these symbolic objects, such as beads, from the 'Moderns'. Which begs the question: why would creatures that have no concept of symbolism scavenge symbolic artefacts? To eat them? This shows the accommodative nature of such mythical constructs of some archaeologists, but it also shows that most other practitioners seem unable to see through such thin arguments.

## The Hobbit myth

One of the most recent examples of a significant blunder has not yet been resolved, so what I present here cannot be a final account. This is a race still being run, and it may be some decades before a clear outcome can be determined. But here is the story so far, followed by a prediction of what it may lead to. One of the key attributes of a scientific proposition is that it must be testable; it must be open to processes of falsification. My proposition about the outcome of this controversy is testable: if we are patient enough and still alive in thirty or forty years, we may witness its closure one day. From Chapter 4 it will have become evident to the reader that it takes that long to resolve any major controversy in paleoanthropology. This is not a coincidence; it is attributable to the general need of the protagonists falling off their perches before a new paradigm can establish itself.

During the early to mid-1990s, in the course of examining the epistemology of the replacement theory, I realised that archaeologists were generally unaware of much of the relevant information. One example of this is the fact that Theodor Verhoeven, a non-archaeologist who had investigated the past of the Indonesian island of Flores in the 1950s and 1960s, had demonstrated that humans occupied this island during an early phase of the Pleistocene (Verhoeven 1958a, 1958b, 1958c, 1968; Maringer 1978; Maringer and Verhoeven 1970a, 1970b, 1970c, 1972, 1977). Just like Boucher de Perthes a good century earlier, he had excavated Lower Palaeolithic stone tools occurring together with the remains of extinct fauna, in his case dominated by stegodons. Since Flores was never connected to the Asian landmass, this meant that the hominins concerned must have crossed the sea, an ability the replacement advocates could not credit these early people with. After writing a couple of rather critical papers on this topic, I prompted a colleague, Mike Morwood, to reopen Verhoeven's excavations and test his claims. He found them fully justified and fine-tuned the dating of the finds to over 800,000 years. *Homo erectus* had obviously reached Flores, and no doubt other Indonesian islands. In 1997 I visited Professor Morwood at his excavations and sensed that, despite his success he was looking for more than stone tools; he was trying to secure skeletal remains of the people concerned. I explained that for this, he needed to dig in a very different kind of site: one with high-pH sediments and in a well-sheltered location, rather than an open site with acidic soils. He asked what I recommended, and I suggested he look for a very spacious limestone cave with a very deep sediment. He did precisely that and ended up excavating in Liang Bua ('Cold Cave'), a cave Verhoeven had investigated much earlier.

In 2004 Dr Morwood and his colleagues reported a new hominin species from the cave, which they named *Homo floresiensis*. Called the 'Hobbit' because of its exceedingly small size, at 1.06 m or so height, it became an immediate world sensation (Figure 41). With a brain size little greater than that of chimps, the little people from Flores were claimed to be so different from any other known hominin that they represented a separate species, if not genus. They were reported to have lived on Flores in the last ten millennia of the Pleistocene, i.e. at a time when the rest of the world was entirely occupied by 'anatomically modern humans', *Homo sapiens sapiens*. This announcement was soon followed by numerous opposing interpretations, and within a year or two, these included practically every possibility from a gibbon-like creature (according to Gert van den Bergh) to a modern but microcephalic human. Some commentators saw the creature as a dwarf *H. erectus*, others as deriving from *H. georgicus* or *H. habilis*. Others again suggested it was an Asian australopithecine, but the strongest argued case was for a severely dwarfed and perhaps pathological modern human. Together with a number of fragmentary specimens, stone tools had also been excavated, and there was no doubt that these were similar to the lithics modern humans made elsewhere about that time. The controversy has raged ever since. The elder statesman of Indonesian paleoanthropology, the late Teuku Jacob (Figure 42), pronounced the fossils as those of insular dwarfs, at least one of which suffered from primary microcephaly, microcrany, microencephaly and micrognathy, which caused mental retardation, and a disruption in brain growth especially in the forehead and cerebellum. This, he said, resulted in a passing resemblance to *H. erectus* and *H. ergaster* skulls.

A bitter confrontation between him and the discoverers of the remains ensued, with many others throughout the world taking sides. The range of views expressed suggests that the discipline is incapable of conclusively deciding whether this is an ape or a modern human, or any possible contender intermediate between these two extremes. Physical anthropologist Maciej Henneberg announced in April 2008 that the first lower left molar of one of the fossils showed signs of dental work. Henneberg detected the presence of material he suggests might be a tooth filling of a type used in the

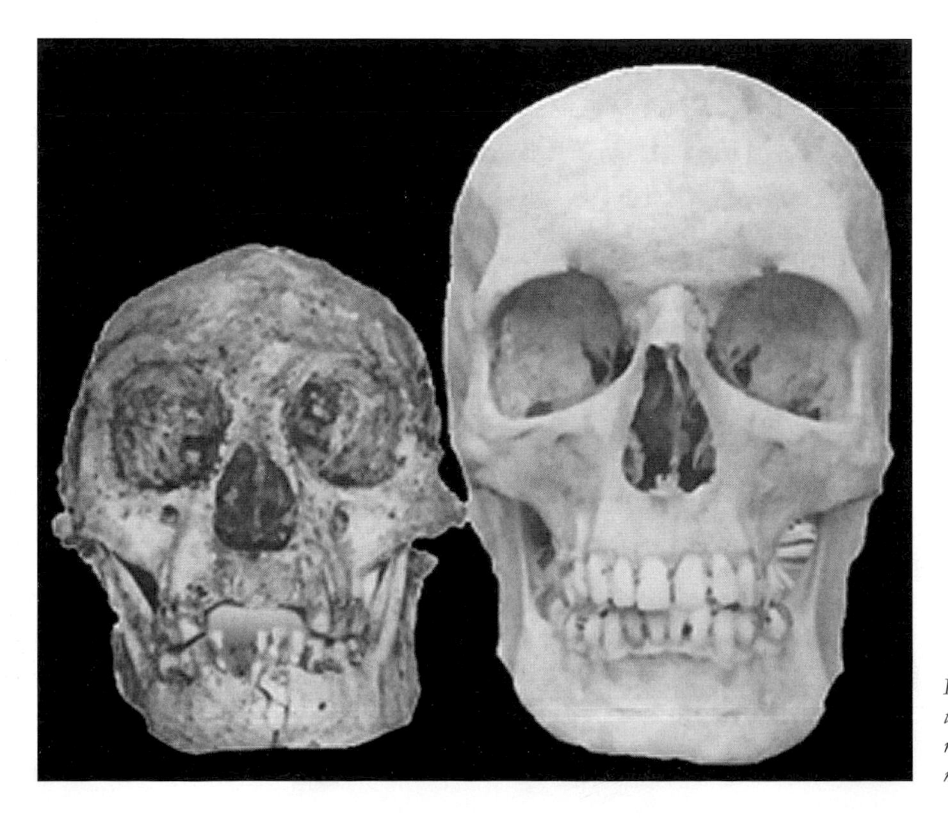

*Figure 41. The skull of the first Flores 'Hobbit' reported, on left, besides a modern skull.*

*Figure 42. Teuku Jacob in 1992, pointing out where the skull of the Mojokerto child was supposedly found. Photograph by C. C. Swisher.*

region around the 1930s (Henneberg and Schofield 2008). The specimen has been claimed to be about 18,000 years old but remains undated, and it is rather difficult to see (though not entirely impossible) how it could have ended up several meters below ground in less than a century. So this would imply a hoax.

It only served to deepen the academic chasm between those experts who see the 'Hobbit' as a distinctive species of very ancient features, and those who see it as a result of severe congenital or genetic defects caused by insular isolation of a tiny gene pool, through extreme genetic drift (Figure 43). As early as 18 February 2005 I commented publicly that "until the [Australian] team considers in its publications ALL Asian pygmy remains of the Pleistocene in a balanced and scientific fashion, their inadequate knowledge of the issue remains a problem." Most importantly the team had not known about the occurrence of other very small humans in the Asia-Pacific region, which is often the result of isolation of small founder populations. For instance they were unaware of the Liang Toge skull, also from Flores, which Verhoeven had considered to be of a proto-Negrito, and of extant Flores pygmy populations (e.g. Rampasasa village). They were unfamiliar with other such specimens, most importantly with the long-established presence of very small people, about the same size as their 'Hobbit', in central India around 200,000 years ago (Bednarik et al. 2005). Announcing to the world a new, 'radically different' species of human in the absence of such relevant knowledge was premature. Among the insular pygmoid populations of the wider region are those found on the Andamans (Sreenathan et al. 2008), whose art I have studied (Figure 44), and on Micronesian islands such as the Palau Archipelago. South African Lee R. Berger excavated with his team Ucheliungs and Omedokel Caves (Chelechol ra Orrak) on the Rock Islands, where at least ten burial caves are known, finding the remains of dozens of tiny human skeletons. These are about the same size as the first Flores specimen, with adult body weight estimated as low as 28 kg, and they, too, exhibit distinctive traits often interpreted as primitive. These include reduction of the absolute size of the face, pronounced supraorbital tori, non-projecting chins, relative megadontia, expansion of the occlusal surface of the premolars, rotation of teeth within the maxilla and mandible, and dental agenesis (Berger et al. 2008). The brain size is not as low as that of the first Flores specimen, but resembles that of *H. erectus*. The Palauen pygmies date from between 2900 and 1400 years ago, after which time they appear to have been replaced by larger people. Berger's team has no hesitation defining them as fully modern *H. sapiens sapiens*, subjected to rapid reduction in body and craniofacial size, echoing my first comment: "Apprehending the full nature of regional variation in Austromelanesian and Pacific Island populations is essential to interpreting the taxonomic status and phylogenetic history of *H. floresiensis*." Therefore the 'Hobbit' may well also be the result of a combination of founder effects, genetic isolation and a high inbreeding coefficient, perhaps manifested as microcephalic osteodysplastic primordial dwarfism. Laron Syndrome is a congenital deficiency of insulin-like growth factor (IGF-I), resulting from inbred genetic defects of the growth hormone (GH) receptor gene (Hershkovitz et al. 2007). Such patients manifest a complete block of the GH-Rs, resulting in a shorter stature than pygmies, who have only a partial defect in the GH receptors. Laron Syndrome yields even lower body heights in adult females than that of the 'Hobbit', the lowest recorded being 95 cm, and in practically all other respects matches its skeletal morphology.

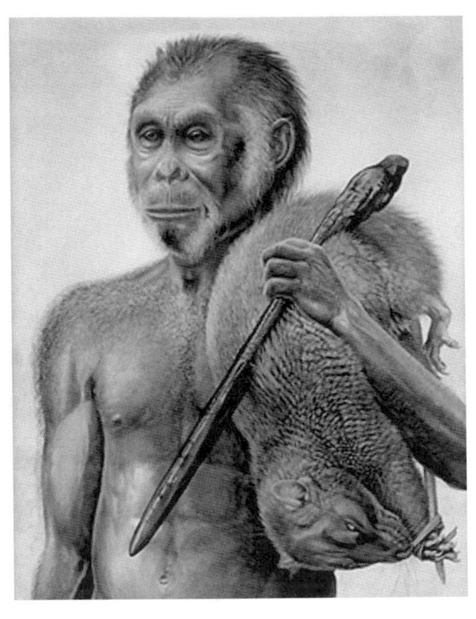

*Figure 43. Artist's rather fanciful reconstruction of the Flores 'Hobbit'.*

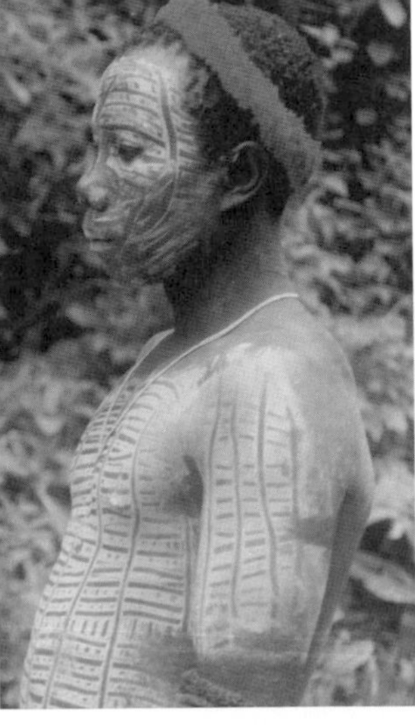

*Figure 44. Pygmoid Jarawa man, Andaman Islands.*

Rapid body size reduction is a common response in endemic island populations, and is certainly not limited to humans: elephants, rhinos and hippos may become as small as pigs, and deer may assume the size of hares. In humans, morphological features considered primitive for the genus (e.g. small brain size, enlarged supraorbital tori, absence of chins, relative megadontia) may be developmental correlates of small body size in pygmoid populations. Even if further specimens of such features were reported from Liang Bua it would not confirm the taxonomic validity of *H. floresiensis*. It is my prediction that in thirty or forty years from now, that 'species' will be included on that ever-growing list of blunders in archaeology. And this is a falsifiable proposition.

## Conclusions

The above six examples of major mistakes in archaeology were selected on the basis of providing the greatest potential variety of different scenarios as possible in the smallest number of examples, and these can be extended by the additional examples I have described in the previous chapter. Thousands upon thousands of mistakes are made by archaeology every year, but most of them are of very minor consequences. Those of interest in the present context are the major controversies that can illuminate the epistemology of the discipline, by showing why and how things went wrong on a spectacular scale.

Therefore each of the examples cited presents us with a different set of circumstances and consequences. It is, however, also evident, on close examination, that a number of consistent themes soon become apparent. I draw attention to the following:

1. The mass media may be manipulated by archaeologists presenting false claims, but sometimes it actually helps exposing bungles, as in the Jinmium or Côa cases (although essentially out of self-interest, at least in respect of Jinmium). As a discipline that produces nothing of economic worth, archaeology needs the patronage of the public, and thus the support of the mass media. Moreover, media coverage affects academic funding of individuals, teams and projects. However, competition for media space is fierce. As a result, stories tend to be embellished for media consumption and researchers can become trapped in the creations they conjured up to draw public attention to their work. This can easily become a factor in having to defend untenable positions.
2. Archaeology, as we have seen in Chapter 3, has a political role, and as a discipline it is also ruled by political considerations. These include loyalty or opposition to colleagues on principle, formation of cliques, exclusion of amateurs and energetic suppression of dissidents. Political dimensions may even include jingoism, as evident in the

'Hobbit' case and others.

3. A generic observation is that many archaeologists have an inadequate comprehension or knowledge of their chosen discipline, and that this does not necessarily prevent them from rising within its hierarchy. This runs like a red thread through even the most diverse examples: we see the shortcomings of much of the entire discipline in the Piltdown, Côa and African Eve controversies (and in all those considered in Chapter 4). Even the remaining cases involve easily identifiable deficiencies in archaeological knowledge, rendering this factor universal to the issue.
4. This leads to a related observation. For better or for worse, English has become the international language of research, and it is widely but falsely assumed that all important archaeological information is available in that language, and has been published prominently. Since all other material is neglected there is a scramble to get one's work published in one of the foremost British or American journals. Yet these journals and their referees tend to adhere to narrow views, resulting in 'academic inbreeding' and an exclusion of alternative views or information not published in English. More than 80% of all available archaeological information has never appeared in English, and much of what is available in English is not covered in the favoured venues.
5. Placing the reputation of the discipline or of individual high-ranking academics above veracity, as we observed in the Glozel and Côa cases — but which is also present in more latent forms in all others — is a key factor in archaeological blunders. That is to be expected, and it is a significant impediment in a discipline comprising unfalsifiable propositions that rely on that factor, the lack of refutability, to uphold capricious notions.

The fear of damage to the discipline through exposing the mistakes of senior academics is perhaps directly related to the weak epistemology of archaeology. To compensate for its lack of refutability, the discipline is very conservative, and it relies heavily on authority. That authority resides primarily in its high-ranking leaders, and any damage to their standing tends to be regarded as damage to the discipline. In other words, the strength of the discipline is perhaps grounded in the 'infallibility' of its elders, a defiance of which may be perceived as challenging the discipline's authority. If one adds to this state the tradition of Western academe, of encouraging the individual to vigorously oppose any contradiction, to defend one's hypotheses at all costs, and recalls that archaeological hypotheses are not falsifiable by archaeological means, two effects are likely to be encountered. Firstly, unfalsifiable propositions are easy to defend by one simple principle: never concede anything. So there is already an unhealthy epistemology inherent at the best of times. Secondly, if one adds to this the perception that the discipline's public prestige would suffer from exposure of mistakes, or from the disclosure that its leaders have erred, the epistemological issue becomes obvious. It is that the discipline would rather preserve its position than concede an error; it prefers dogma to veracity.

At that point it ceases to be a scholarly pursuit, i.e. one existing purely for the sake of finding truth. It has become a belief system and a hegemony.

## REFERENCES

Bednarik, R. G. 1993. Palaeolithic art in India. *Man and Environment* 18(2): 33–40.

Bednarik, R. G. 2007. Antiquity and authorship of the Chauvet rock art. *Rock Art Research* 24(1): 21–34.

Bednarik, R. G. 2008a. The mythical Moderns. *Journal of World Prehistory* 21(2): 85–102.

Bednarik, R. G. 2008b. The domestication of humans. *Anthropologie* 46(1): 1–17.

Bednarik, R. G. 2009a. Fluvial erosion of inscriptions and petroglyphs at Siega Verde, Spain. *Journal of Archaeological Science* 36(10): 2365–2373.

Bednarik, R. G. 2009b. To be or not to be Palaeolithic, that is the question. *Rock Art Research* 26(2): 165–177.

Bednarik, R. G. 2011. *The human condition*. Springer, New York.

Bednarik, R. G., G. Kumar, A. Watchman and R. G. Roberts 2005. Preliminary results of the EIP Project. *Rock Art Research* 22(2): 147–197.

Berger, L. R., S. E. Churchill, B. De Klerk and R. L. Quinn 2008. Small-bodied humans from Palau, Micronesia. *PLoS ONE* 3(3): e1780. DOI:*10.1371/journal.pone.0001780*.

Clermont, N. 1992. On the Piltdown joker and accomplice: a French connection? *Current Anthropology* 33: 587.

Dart R. A. 1925. *Australopithecus africanus*: the man-ape of South Africa. *Nature* 115: 195–199.

Gardiner, B. G. 2003. The Piltdown forgery: a re-statement of the case against Hinton. *Zoological Journal of the Linnean Society* 139: 315–335.

Green, R. E., J. Krause, A. W. Briggs, T. Maricic, U. Stenzel, M. Kircher, N. Patterson, H. Li, W. Zhai, M. Hsi-Yang Fritz, N. F. Hansen, E. Y. Durand, A.-S. Malaspinas, J. D. Jensen, T. Marques-Bonet, C. Alkan, K. Prüfer, M. Meyer, H. A. Burbano, J. M. Good, R. Schultz, A. Aximu-Petri, A. Butthof, B. Höber, B. Höffner, M. Siegemund, A. Weihmann, C. Nusbaum, E. S. Lander, C. Russ, N. Novod, J. Affourtit, M. Egholm, C. Verna, P. Rudan, D. Brajkovic, Ž. Kucan, I. Gušic, V. B. Doronichev, L. V. Golovanova, C. Lalueza-Fox, M. de la Rasilla, J. Fortea, A. Rosas, R. W. Schmitz, P. L. F. Johnson, E. E. Eichler, D. Falush, E. Birney, J. C. Mullikin, M. Slatkin, R. Nielsen, J. Kelso, M. Lachmann, D. Reich and S. Pääbo 2010. A draft sequence of the Neandertal genome. *Science* 328: 710–722.

Henneberg, M. and J. Schofield 2008. *The Hobbit trap: money, fame, science and the discovery of a 'new species'*. Wakefield Press, Kent Town, Australia.

Hershkovitz, I., L. Kornreichand and Z. Laron 2007. Comparative skeletal features between *Homo floresiensis* and patients with primary growth hormone insensitivity (Laron syndrome). *American Journal of Physical Anthropology* 134 (2): 198–208.

Krause, J., Q. Fu, J. M. Good, B. Viola, M. V. Shunkov, A. P. Derevianko and S. Pääbo 2010. The complete mitochondrial DNA genome of an unknown hominin from southern Siberia. *Nature* 464(7290): 894–897.

Maringer, J. 1978. Ein paläolithischer Schaber aus gelbgeädertem schwarzem Opal (Flores, Indonesien). *Anthropos* 73: 597.

Maringer, J. and T. Verhoeven 1970a. Die Steinartefakte aus der Stegodon-Fossilschicht von Mengeruda auf Flores, Indonesien. *Anthropos* 65: 229–247.

Maringer, J. and T. Verhoeven 1970b. Note on some stone artifacts in the National Archaeological Institute of Indonesia at Djakarta, collected from the Stegodon-fossil bed at Boaleza in Flores. *Anthropos* 65: 638–639.

Maringer, J. and T. Verhoeven 1970c. Die Oberflächenfunde aus dem Fossilgebiet von Mengeruda und Olabula auf Flores, Indonesien. *Anthropos* 65: 530–546.

Maringer, J. and T. Verhoeven 1972. Steingeräte aus dem Waiklau-Trockenbett bei Maumere auf Flores, Indonesien. Eine Patjitanian-artige Industrie auf der Insel Flores. *Anthropos* 67: 129–137.

Maringer, J. and T. Verhoeven 1975. Die Oberflächenfunde von Marokoak auf Flores, Indonesien. Ein weiterer altpaläolithischer Fundkomplex von Flores. *Anthropos* 70: 97–104.

Maringer, J. and T. Verhoeven 1977. Ein paläolithischer Höhlenfundplatz auf der Insel Flores, Indonesien. *Anthropos* 72: 256–273.

Reich, D., R. E. Green, M. Kircher, J. Krause, N. Patterson, E. Y. Durand, B. Viola, A. W. Briggs, U. Stenzel, P. L. F. Johnson, T. Maricic, J. M. Good, T. Marques-Bonet, C. Alkan, Q. Fu, S. Mallick, H. Li, M. Meyer, E. E. Eichler, M. Stoneking, M. Richards, S. Talamo, M. V. Shunkov, A. P. Derevianko, J.-J. Hublin, J. Kelso, M. Slatkin and S. Pääbo 2010. Genetic history of an archaic hominin group from Denisova Cave in Siberia. *Nature, 23 December 2010, doi:10.1038/nature09710.*

Roberts, R. M. Bird, J. Olley, R. Galbraith, E. Lawson, G. Laslett, H. Yoshida, R. Jones, R. Fullagar, G. Jacobsen and Q. Hua 1998. Optical and radiocarbon dating at Jinmium rock shelter in northern Australia. Nature 393: 358–362.

Rothwell, N. 1996a. Politics etched in stone. *The Australian*, 23 September, pp. 1 and 4.

Rothwell, N. 1996b. Scientists split over rock find's implications for evolution: challenge to the origin of man. *The Australian*, 23 September, pp. 1 and 5.

Rothwell, N. 1996c. Our origins on rock ground. *The Australian*, 28–29 September, p. 27.

Schulz, M. 2004. Die Regeln mache ich. *Der Spiegel* 34(18 August): 128–131.

Sreenathan, M., V. R. Rao and R. G. Bednarik 2008. Palaeolithic cognitive inheritance in aesthetic behavior of the Jarawas of the Andaman Islands. *Anthropos* 103: 367–392.

Thackeray, F. 1992. On the Piltdown joker and accomplice: a French connection? *Current Anthropology* 33: 587–589.

Teilhard de Chardin, P. 1913. La prehistoire et ses progress. *Études* 1913: 40–53.

Verhoeven, T. 1958a. Pleistozäne Funde in Flores. *Anthropos* 53: 264–265.

Verhoeven, T. 1958b. Proto-Negrito in den Grotten auf Flores. *Anthropos* 53: 229–232.

Verhoeven, T. 1958c. Neue Funde prähistorischer Fauna in Flores. *Anthropos* 53: 262–263.

Verhoeven, T. 1968. Vorgeschichtliche Forschungen auf Flores, Timor und Sumba. In *Anthropica: Gedenkschrift zum 100. Geburtstag von P. W. Schmidt*, pp. 393–403. Studia Instituti Anthropos No. 21, St. Augustin.

Weiner, W. S., K. P. Oakley, W. E. Le Gros Clark 1953. The solution of the Piltdown problem. *Bulletin of the British Museum (Natural History) Geology* 2(3): 141–146.

# 6

# LOGIC IN ARCHAEOLOGY

## Introduction

The first five chapters of this book could easily create the impression that its purpose is to discredit archaeology as a discipline. Nothing could be further from the truth. Their purpose was to explain epistemological impairments of various types, in an effort to understand contingent theoretical and procedural deficiencies that are amenable to correction. Having thus clarified specific factors contributing to the epistemic malaise of archaeology, practical as well as theoretical, it is high time to demonstrate a crucial justification for such a critique: do I have a better alternative to offer? I will present such an alternative approach in this chapter, and expand on it in the next.

This discussion is primarily about Pleistocene archaeology, because Holocene archaeology, covering the last 10,500 years or so, is in significantly better shape. Indeed, the closer we come to the present time, the more secure our models seem to be. The archaeology of the medieval and more recent times appears to be little more than a filling in of minor lacunae of historical reconstruction. To a perhaps lesser degree that can also be said about the Roman, Greek and even earlier Classical periods and societies. However, by the time we explore beyond the introduction of writing, into the early agrarian cultures of the mid- to late-Holocene, the veracity of archaeological explanations and interpretations tends to become increasingly hazy, because it depends more and more on authority and epistemological frameworks as we proceed further back into time. Significant deficiencies appear as we probe into the period claimed to mark the advent of semi-sedentary societies, named the Mesolithic in Europe. Here we find the first suggestions of major misinterpretation of the empirical evidence.

This point is readily demonstrated. The Mesolithic or Middle Stone Age sits rather uneasily between two arbitrarily defined major 'cultural' eras: the farming communities of the 'Neolithic' time, and the presumably nomadic peoples of the 'Upper Palaeolithic' periods. Not only is this supposed intermediate stage in human development rather vaguely defined in western Europe, it seems to be largely lacking in south-eastern Europe, northern Africa and in the important region of south-western Asia. Indeed, the term has really little meaning in any other continent. In western Europe it defines

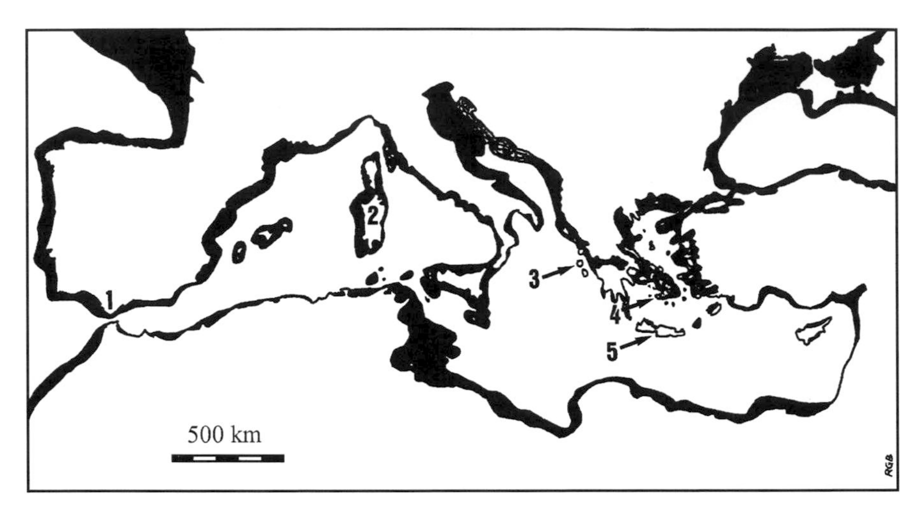


*Figure 45. Former Pleistocene sea levels in the Mediterranean, showing the locations of water crossings of the Pleistocene.*

societies that may or may not have had semi-permanent settlements, that focused on specific food animals (which simply marks a response to climatic and ecological changes), used microliths (as did previous cultures for tens of millennia) and interment practices (ditto), and the only distinctive feature is the sudden appearance of coastal adaptation systems, including great shellfish middens. This is an entirely meaningless criterion, because we have not the faintest idea about the coastal cultures of the preceding Upper Palaeolithic. Rising sea levels towards the end of the Pleistocene and in the first millennia of the Holocene ensured that all earlier evidence below about 140 m above the former sea level was obliterated (Figure 45). Indeed, throughout the Pleistocene, the sea rose and fell many times, which practically halves the range of potentially recoverable evidence about the entire period in a very systematic fashion. The low-lying river deltas, the coastal zones and the lower parts of major valleys always presented the richest environments and would have always attracted far greater population densities, as indeed they still do today: over half the world's population resides in regions that would be submerged by the kinds of sea level rises experienced in the Pleistocene. It is also evident that, in contrast to the tribes of the hinterland, of the highlands, steppes and jungles, the coastal tribes would have been far more sedentary. But, alas, we have no knowledge of any kind about the ethnicity, technology, culture, or way of life of any Pleistocene people who lived in these favourable environments, anywhere in the world. What we do know is this: as the sea level rose to roughly its present level, we see the sudden appearance of evidence of coastal societies we define as Epipalaeolithic or Mesolithic. It is far more logical to see in them the coastal counterpart to the hinterland version of late Upper Palaeolithic economies. Indeed, it would be strange if coastal adaptations had not been pushed to what is now dry land as the sea rose. But instead of this logical explanation, European archaeology has opted for an implausible alternative, inventing a cultural phase.

One may object that, while plausible, there is by definition no empirical evidence for this proposition. But there is, if one looks for it. All evidence at the seacoasts was of course destroyed, but there may be relevant information from the littoral zones of large lakes that, due to falling aquifer levels, disappeared during the Pleistocene. Lake Fezzan in the Libyan Sahara is such a case, and here we have evidence not only of exceptional cultural sophistication by about 400,000 years ago (Ziegert 2007, 2010), we also have navigation evidence (Bednarik 1999a; Werry and Kazenwadel 1999), as explained below. These rare finds tell us far more about the development of hominin culture than any others on the face of the earth, but they are systematically ignored because they contradict mainstream dogma.

The importance of this example is that it shows that there are variables in the archaeology of early phases that can dramatically distort the evidence because of their systematic effects. The sea level changes have affected many other aspects of the surviving record. For instance we know that people of the Pleistocene must have crossed the sea on numerous occasions, because they colonised territories that were never accessible by land, or at least not during the existence of humans. But we cannot reasonably expect to find any remains of the vessels they used in these adventures. Because of the sea level rise around 9000 years ago, the oldest watercraft ever found, in Holland, is precisely that age (Figure 46).

But there are many other factors that determine the survival of evidence in systematic patterns, so it is fundamentally false to draw one-to-one deductions from the archaeological evidence. Rather there needs to be a program, a very deliberate effort, of making allowances for all the systematic distortions likely to affect empirical data. Such a program does exist, it has been proposed in the early 1990s, but remains entirely ignored by mainstream archaeology. It is called *taphonomic logic.*

## The role of taphonomy

Distinguished by its frequent absence in dictionaries (even those that normally do contain important terms), the word *taphonomy* referred initially to the study of the processes to which preserved organic remains had been subjected. Hence the word's original use was in palaeontology, but even here the concept has only been seriously considered since the 1960s. Efremov (1940) introduced the term in an effort to seek laws explaining the processes relating

*Figure 46. The dugout canoe from Pesse, Holland, about 8300 years old.*

to the burial of bones within a single framework. Palaeontologists gradually took up this fundamentally scientific approach to the study of fossil remains over the following decades (e.g. Behrensmeyer 1975; 1978; Gifford 1981; Hill 1976, 1979). During the 1980s, archaeologists realised that the underlying principles also applied to their discipline, particularly to the organic aspects of site formation processes (Brain 1981; Binford 1981). After initially restricting their application mostly to faunal remains, some scholars eventually perceived that the concept has broader applications. In Australia, Hiscock realised that the underlying principles were also applicable to stone implements (Hiscock 1985, 1990).

Today the word taphonomy as it applies in archaeology has become somewhat of a misnomer: *tapho-* is Greek for grave, and *-nomy* indicates systematisation of knowledge. Archaeologically, taphonomy is now taken to refer to the study of the transformation of materials into the 'archaeological record' (Bahn 1992: 489). It has become evident that even this expanded definition may not be adequate, and particularly, that a taphonomy of palaeoart, including rock art, places more rigorous demands on practitioners. But to explain this it is useful first to consider the palaeontological application of taphonomic concepts, to see what can be gleaned from them. That these experiences can be usefully employed in archaeology and rock art research only shows what a powerful epistemological tool taphonomic logic (Bednarik 1990–91, 1992, 1993a, 1994) is.

In palaeontology, taphonomy covers all events during the transition of animal and plant remains from the biosphere to the lithosphere, including mode of death, scavenging, ingestion and digestive processes, transport (by animals, wind, water or sediment movement), surface weathering and geological erosion, trampling, differential dissolution of tissues and mineralisation or other replacement processes, and even modification of osteal remains as tools by hominins. The organic remains one recovers bear forensic evidence of their preservational history, including degree of completeness, damage patterns, orientation in respect to other debris, surface wear and alteration results. Without a good understanding of how these many processes may have affected statistical indices of the material it would be fairly futile to reconstruct biological models of the species or, indeed, the ecosystem in question.

This form of taphonomy aims to elucidate how biological information has been altered from the original living systems to a 'fossil record', by biological, physical and chemical degradation or alteration processes. To consider a specific example: the differences in the distribution of individuals in a present environment and one implied from fossil evidence relating to a similar kind can differ most dramatically. The Pleistocene cave bear (*Ursus spelaeus* Rosenmüller and Heinroth) is so named because over 99% of its remains were found in caves. They have been recovered in massive numbers from the sediments of cave lairs, representing many tens of thousands of individuals in some caves (Bednarik 1993b). For instance some 250 tonnes of cave

*Figure 47. The massive deposit of cave bear bones in the Drachenhöhle of Styria, Austria.*

bear bones were excavated from the Drachenhöhle in Austria alone (Figure 47). Therefore the distribution pattern of these remains would suggest that the animal was a habitual cave dweller that normally died inside caves. Yet the species only sought out caves as winter hibernation sites during stadial periods (the peaks of the Ice Ages), otherwise it spent no time at all in caves. It was primarily a herbivore whose principal diet consisted of grass. Weakened or old bears often died during hibernation, but there can be no doubt that the vast majority of the cave bears died outside of caves. So why are their remains found almost exclusively in caves?

The answer lies in taphonomy: the probability of the skeletal remains surviving in cave sediments is thousands of times greater than that of surviving in the open. There are essentially two reasons for this: the sedimentary pH, which determines the survivability of osteal remains, is very high inside limestone caves; and they are not subjected to external weathering, but to very stable speleoclimatic conditions. Similar principles apply throughout archaeology, but are widely misunderstood by archaeologists. No data of the distribution of a phenomenon in the Pleistocene can be meaningful without considering such principles. These underlying axioms are not fully exploited by taphonomy itself, or only so for a very specific purpose: to determine what one had to consider to deduce aspects of a living system from the fossil aspects that survive from it. But the underlying principle, like the law of uniformitarianism, is much more profound, and can be defined more broadly, and universally applicable. It can be formulated thus: *the surviving and humanly detectable traces of a phenomenon of the past can define that phenomenon only to the extent that adequate recourse to a specific form of logic has been applied*; it is called 'taphonomic logic' (Bednarik 1990–91, 1994). It applies to any phenomenon or event of the past, be it astronomical, sedimentary, geological, palaeontological, palynological or archaeological. This form of logic is quantifiable (at least as an integral function; Bednarik 1994: 73) and is not a hypothesis presented for testing, but a theorem facilitating the assessment of past conditions that cannot be observed directly. This has the most profound effects on our ability to interpret what we regard as archaeological evidence.

It is unfortunately the case that so far, most archaeologists have not understood the relevance of this epistemological tool to their discipline (as noted by Soloman 1990). Contrary to their common perception, taphonomy does not inherently deal with osteal remains (bones and teeth). It could be seen as pure coincidence that the underlying principle was first identified in palaeontology. In essence, taphonomy deals with the logic underpinning the idea that the quantified characteristics of a record of past events or systems are not an accurate reflection of what would have been a record of the live system or observed event. It follows that to gain any level of valid understanding of the 'live system' (in the broadest and most inclusive sense) one has to explore the processes that led to the extant traces.

The probably greatest single epistemological encumbrance of archaeology as it has been conducted is the tendency of treating 'empirical evidence' as representing a random sample — as if it amounted to a representative selection of variables defining the entity being explored. The concept of 'random sample' is taken from the practices of the hard sciences, where it is crucial that it is in fact representative. In archaeology, however, it is impossible to secure samples of culture that can be representative of any condition: each site deposit, and each part of each site, is culturally unique. Therefore uniformitarianist principles, which apply to other species and systems, do not apply to the cultural systems of humans. Very simply, the logic of mainstream archaeology in explaining evidence must be inherently false and misleading. Representativeness is manufactured by the archaeologist, who arranges series of objects arbitrarily and creates taxonomy from them, then attributing them to 'cultures'. To illustrate how this process can lead to absurd systems, we may consider its application to rock art. Major rock art sites are almost always cumulative assemblages in generally two-dimensional space (ignoring here the possibility of detecting nano-stratigraphies, a technique introduced in the 1970s; Bednarik 1979). The scientific dating of these sequences remains extremely difficult (Bednarik 2002a). So we have single sites or rock panels bearing the artistic precipitate of different periods, perhaps different cultures (we will consider this issue in more detail in the next chapter).

Treating these accumulations as deriving from single cultures can only lead to falsities, and there are countless examples in archaeology where this has taken place. In epistemological parlance, there is a dependency relation called a supervenience: one set of properties (forming a historical event) is supervenient on a second set (represented in the selected sample). The relationship between the two sets cannot therefore be explored by traditional deductive reasoning. However, even if one made allowances for the purely taphonomic issues (the enormous variations in the survival rates of different classes of evidence), the disparities would not be solved (see below). These variations are much greater than most practitioners realise. Of all the events that occurred during the archaeological past, no evidence of any kind survived for more than a second in 99.999% of all cases. Of the still innumerable remaining instances, evidence survives to this day only in a tiny fraction of one-millionth of a percent. Of this remaining 'sample', only an infinitesimal portion can reasonably be assumed to have been recovered, of which an even smaller part has been correctly interpreted. This introduces an even more profound issue: not only do we need to understand the systematic biases of preservation; we also need to consider those of recovery and interpretation.

The full complexity of the issue is perhaps best illustrated by example. We could consider the many factors that contribute to the relative over-representation of, say, gold objects in the archaeological record. Apart from the obvious advantage in preservation of a noble metal, gold objects are far more likely to be collected, noticed, salvaged, recorded or found with metal detectors than other remains. Moreover, they are more likely to occur in select places — tumuli, shipwrecks, pyramids, or hoards — especially likely to attract the interest of archaeologists, who may well prefer not to dig in places without promise. Even the preoccupations of archaeologists become issues resembling taphonomic factors, and are decisive in determining what we innocently call the 'archaeological record'. Once found, a gold object is more likely to be mentioned in a publication than, say, a bone object. Thus the observation that there are *x* times more bone objects than gold objects in the 'archaeological record', without further qualification, is meaningless. Even a snowman made by a 'Neanderthal' could theoretically have survived, while many gold objects have been destroyed. Therefore probability of survival can never be nil, nor can it be 100%. Or, to use the language of taphonomic logic, gold objects have an extremely short *taphonomic lag time* (effectively the time span between a phenomenon's introduction and its first common appearance on the available 'archaeological record'), snowmen have an extremely long one. The point in time separating the taphonomic lag from the period from which specimens occur in good numbers is called the taphonomic threshold, which must lie somewhere between the phenomenon's first appearance and the present, but can never coincide with either (Figure 48). The importance of this is that, for the vast majority of phenomenon categories, such as objects of leather, cordage, bark and so forth, the lag time tends to account for well over 99% of the duration of their existence in the past. On the other hand, for most phenomenon categories it is perfectly possible for highly isolated instances to occur beyond the threshold. Most archaeological misinterpretation of the past is relatable to a lack of appreciation of these factors, which inevitably leads to minimalist assumptions and endemic under-rating of the societies concerned: perceived absence of evidence is interpreted as evidence of absence, and isolated specimens from the taphonomic lag time are sometimes explained away as "running ahead of time" (Vishnyatski 1999), but are more commonly rejected as flukes, as the result of faulty stratigraphy

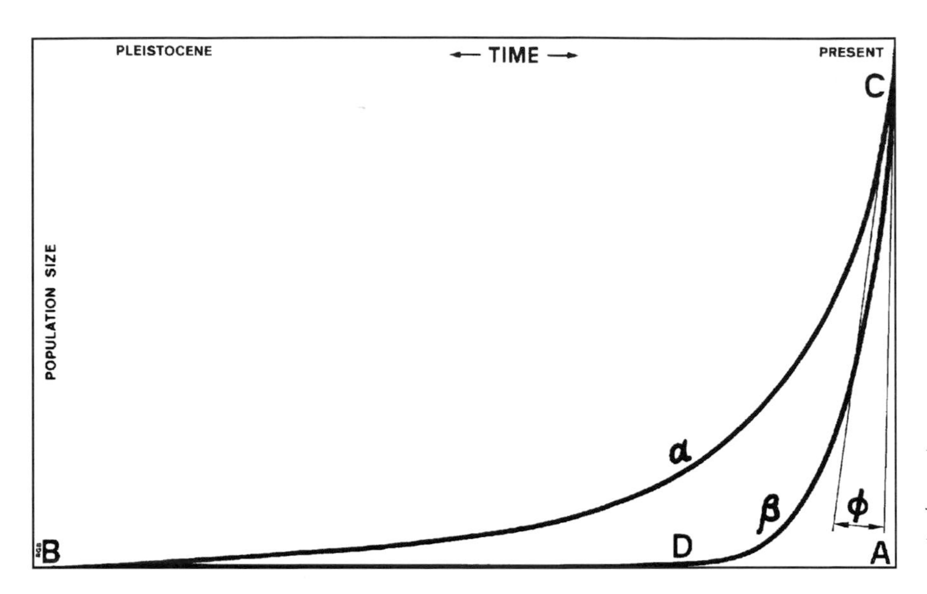


*Figure 48. Principles of the relationship of total production of an archaeological phenomenon $s_\alpha$ to its surviving instances $s_\beta$ as a function of angle φ. These principles are the basis of taphonomic logic.*

and so forth. Therein lies the simple explanation of the huge gap of credibility between dominant paradigms of Pleistocene archaeology and the models demanded by other disciplines, by rational thought and by plain common sense. Typically, the discipline under-interprets significantly the technological, cognitive, intellectual and cultural levels of all hominins of the Pleistocene, and this is the reason for it.

## Metamorphology: an alternative

The observation that a taphonomic logic-style of discourse needs to be applied to various factors other than taphonomy proper expands the scope of this discussion considerably. The solution is the introduction of metamorphology as the scientific version of archaeology. It is a logic-based, refutable system of reviewing archaeological information that determines whether archaeological propositions could have scientific legitimacy. It is developed especially from taphonomic logic, which as we have seen hinges on the concept of cumulative data loss as a function of time (the principle is depicted graphically in Figure 48). It replaces inductive uniformitarianism, hitherto the de-facto basis (Cameron 1993), as a unified theory, of archaeology.

Metamorphology (Bednarik 1995, 2006) is the science of how forms of evidence of events in the past become the forms as which they are perceived or understood by the individual researcher today. In accounting for the considerable gap that exists between the reality of what actually happened at some point of time in the distant past, and the abstraction of it as it is perceived by, for example, an archaeologist, it is crucial to focus on the individual. The discipline is not a quasi-democratic reflection of the view of all (see Chapter 2); its paradigm is based on the authority of a political hegemony. Metamorphology obviously has to take into consideration myriad factors and it cannot be expected to provide precise interpretations, but it needs to determine how the individual interpreter of the past arrives at his or her pronouncements. Knowledge in archaeology is not some mysterious collective unconscious to which practitioners are somehow connected; it is individual knowledge of individual practitioners, limited by many factors. Like refutation in general, metamorphology provides us with models of what is unlikely to be valid, and so strengthens archaeology by weakening its dogmas. It also rejects the concept of an 'archaeological record', or a collective knowledge of the discipline.

The most obvious of these factors accounting for metamorphology, and for the gap between archaeologists' constructs and what really happened in the past, is taphonomy. It distorts archaeological evidence systematically, and it does so in forms that have often not been appreciated adequately. Indeed, after the palaeontological concept of taphonomy was introduced into archaeology over forty years after its 1940 inception, it was soon misunderstood and effectively became actuopalaeontology — which ironically taphonomy was originally intended to replace (see Soloman 1990 for a superb discussion). Hence the potential of taphonomy itself has remained significantly under-utilised in archaeology in more ways than one. But if the inherent principles are extended to the methods of *recovery* of evidence; those of its *interpretation*; those of its *reporting* and selective *dissemination*; those of its *statistical treatment*; or to the individual researcher's own *biases* and *limitations*; and to a variety of other factors, it becomes apparent that these also tend to be systematic. These factors may include the priorities of research traditions, of individual leaders in the discipline, of specific institutions, of funding agencies or of society as a whole. In Chapter 3 and elsewhere we have already met some of these factors of bias. There can be no doubt that there is a very considerable gap between the reality of what happened in the distant past, and the abstraction of it as perceived by the individual archaeologist interpreting a specific, subjectively selected and non-random sample of the remaining evidence. This is already obvious from the observation that there is no universal world archaeology; there are countless archaeologies (see Chapter 2). To account for this gap of incommensurability, to decide what the distorting factors are and what their respective effects and interplay might be, a separate sub-discipline is required that addresses the epistemological basis of archaeology. Taphonomy itself is not the whole answer, because it accounts for only *some* of these truncating and modifying factors.

For metamorphology to be scientific, its propositions must be refutable. It is logic based and draws heavily on knowledge of taphonomic processes, and on a variety of other falsifiable observations. A unified theory of metamorphology has been formulated and published, at least in embryonic form (Bednarik 1995, 2006). It has been shown that metamorphological quantification, although extremely difficult, should be possible, at least in general or abstract forms (e.g. as integral functions). It extends the underlying principle of taphonomic logic (that scientific access to the human past is contingent on the coherent identification of *that part of the extant characteristics of the evidence that is not the result of taphonomic processes*; Bednarik 1990–91) to all aspects of archaeological interpretation. These include the way data are *collected, stored, interpreted and disseminated.* They include the biases of the individual researcher (cognitive, religious, ontological, academic, intellectual), of specific schools, or the discipline as a whole, and many other external factors that have a bearing on how the so-called evidence is individually perceived, reported and interpreted. For instance, the researcher's own limitations are a powerful factor in how evidence may be reported. These may be limitations of knowledge or of language. Ignorance of researchers concerning existing data, language barriers, and biases through preconceived models has not only severely influenced hypotheses and their defence; it has also stifled the flow of information in palaeoart studies and archaeology (e.g. Bednarik 1992, 1995b, 1995c, 1999b). It is certainly a quantifiable factor. The academic system itself, which is so crucial to the dissemination of knowledge, can also stifle that very process and act as a filter in quite a number of ways. All of this can cumulatively add up to such distortions in dominant models that these bear little resemblance to what historically happened. This is because many of the distortions are not random; *they are systematic.* As a result, the human past as it is perceived at present is merely a constructed artefact, which is the central thesis of this book. It has very limited scientific credibility.

To correct this we need to be able to understand the nature and effects of these distortions, be they taphonomic or related to other epistemic encumbrances. This would provide the kind of framework we require to account for the gap between what happened in the distant past, and the abstraction or reified construct of it as it is perceived by the individual researcher interpreting a specific 'sample' of the remaining evidence of this event or connected events. For instance, we need to understand the effects of false hypotheses and of their ardent defence if we are to obtain a valid reflection of metamorphology. There is surely no reason why the dynamics of knowledge acquisition or academic power politics in the discipline should be immune from scholarly analysis. Archaeology, like anthropology, often does not hesitate to study the taboos of the societies it investigates, so the study of itself should not be taboo either. These are realities, they have significant effects on the discipline, and these dynamics need to be understood like any other process contributing to our knowledge. Therefore this aspect should be studied as carefully as any other that contributes to metamorphology. The discipline would be in a sorry state if such research would be discouraged because the 'reputation' or sensitivities of individuals are considered to have precedence over its integrity or veracity.

## Why the dominant paradigm is wrong

All these considerations are, however, theoretical, and to probe their practical applications and implications we return to the specific example we began this subject with: the effects of Pleistocene sea-level changes. The self-evident corollary would be that, to secure a balanced picture of the cultures, technologies and genetics of Pleistocene humans, we would need to take into consideration that part of humanity that lived in the most fertile and resource-rich environments, i.e. those of low elevations. We can reasonably assume that it made up well over half of humanity, perhaps even far more than half, and that it was almost certainly more developed and more sedentary. But with the exception of a few unique glimpses, we know nothing about these people. Pleistocene archaeologists appear uniformly incapable of appreciating the effects of having studied nothing more than the remains of mobile inland tribes that followed the herds of the steppes, upper valleys and highlands. To imply that this is the complete story of Pleistocene people is severely misleading, and demonstrably false. It is precisely the reason why most Pleistocene archaeologists find it hard to explain how Middle Palaeolithic seafarers could have been capable of settling Australia, travelling many days across the open sea to do so (Bednarik and Kuckenburg 1999). Much earlier, almost a million years ago, their Lower Palaeolithic ancestors, of *Homo erectus* stock (Figure 49), reached the islands of Wallacea across the sea (Bednarik 1999b, 2003a).

*Figure 49. Artist's impression of a* Homo erectus *man constructing a bamboo raft.*

Most orthodox archaeologists have great difficulty accepting this, which indicates how severely their thinking has been conditioned by false models. To successfully colonise any landmass for the long term, a founding population of adequate genetic variability to result in a viable breeding group is absolutely essential (we have seen with the Flores 'Hobbit' and other insular populations what happens when a gene pool is too small). This demands that at least dozens, preferably hundreds of people, comprising a good number of fertile females, had to travel, implying considerable organisational and language ability (Figure 24). We lack any knowledge of the coastal tribes of these hominins, but the only logical explanation for their many successful colonisations in Indonesia as well as in the Mediterranean (Bednarik 1999a, 1999b) is that their technology was more advanced than that of the inland tribes. Orthodox archaeology assumes that permanent settlements were introduced with the 'Neolithic revolution', unaware that the earliest-known villages of stone huts date from the Acheulian, 400,000 years earlier (Ziegert 2007). But of course they could not be found in the coastal zones along seashores. They were located on a former giant inland lake in the Libyan Sahara, Fezzan Lake, which was then twice the present size of Lake Victoria. That lake has since disappeared together with its aquifer, which is precisely why the remains could survive in these unusual circumstances. They allow a rare glimpse into the sophistication of a littoral people of the Lower Palaeolithic, showing that complex villages of stone-walled huts, which apparently arrived in inland regions only with the Holocene, have been in use for hundreds of millennia where permanent food sources permitted sedentary settlements. There is even evidence of islands in Fezzan Lake having been reached in the Acheulian, probably by reed rafts (Werry and Kazenwadel 1999). This evidence can explain many aspects: the early appearance of palaeoart, of beads and pendants (found there at 200,000 years BP), of Lower Palaeolithic seafaring, or the ability of hominins to colonise regions of cold climates. It would, however, also imply that the ecology of the Palaeolithic periods has been completely misinterpreted; that the orthodox Pleistocene paradigm is just a monumental distortion of history.

It may seem inconceivable that, after 150 years of research, such a state should still be possible, but when the many errors in the history of this discipline (some of which were listed in Chapters 4 and 5) are considered, the petulant question does arise: have these patterns of the past been left behind? On closer examination, evidence to the contrary does emerge. Consider, for instance, the intricate mythology archaeology has created around the Ice Age cave art of south-western Europe: almost every key interpretive aspect of it appears to be false. Symboling did not, as claimed almost universally, commence with the advent of the Upper Palaeolithic in Europe, but at least twenty times as long ago. Even the traditional Darwinist sequence of emerging symbolic capabilities needs to be discarded. Franco-Cantabrian cave art is not a form of rock art endemic to caves; its exclusive occurrence in deep limestone caves is almost certainly a taphonomic phenomenon (Bednarik 1994). It was not created by shamans or great artists; much or most of it appears to be the work of children and teenagers (Bednarik 1986, 2002b, 2008; Guthrie 2005). Ethnographic evidence suggests that figurative art may by some societies be regarded as a 'juvenile' art form (Sreenathan et al. 2008; Bednarik and Sreenathan 2012). Contrary to a widely held view, figurative imagery is cognitively less developed than non-figurative. Whereas in figurative symbolism, the connection between referent and referrer is purely via iconicity — a relatively simple cognitive factor building on visual ambiguity and accessible even to animals other then humans — the symbolism of non-iconic art is only navigable by possessing the relevant cultural 'software'. Figurative art results from a deliberate creation of visual ambiguity (Bednarik 2003b: 408, 412) and is therefore based on lower levels of perception and neural disambiguation than non-figurative art.

If one adds to these considerations the orthodox misconceptions that European Palaeolithic cave art consists mainly of zoomorphs (it does not; most of it is non-figurative); or that zoomorphs mark Pleistocene art (less than 1% of the world's surviving Ice Age palaeoart is even figurative); or that all Pleistocene rock art is of the Upper Palaeolithic (in fact its lion share is Middle Palaeolithic rock art, especially in Australia and southern Africa); or that the art of the early Upper Palaeolithic (such as that of the Châtelperronian and Aurignacian) is the work of anatomically fully modern humans (it appears to be made by Neanderthaloid people), one begins to appreciate the depth of the issue. Practically *all widely held beliefs about Pleistocene palaeoart are either false, or are very probably false.* This is a discipline in deep denial.

Having shown how many challenges a single taphonomic factor, sea-level fluctuation, provides to the mainstream, orthodox paradigm of Pleistocene human history, I need to emphasise that these challenges can be multiplied hundreds of times, through hundreds of similar factors. This was presented as just one of many examples illustrating the effects of metamorphology: taphonomic logic, combined with a rigorous review of the practices of interpretation and dissemination, can and must be applied to all archaeological claims about Pleistocene humanity. This, put very simply, is the basis of a sound epistemology of Pleistocene archaeology.

## REFERENCES

Bahn, P. (ed.) 1992. *Collins dictionary of archaeology*. HarperCollins Publishers, Glasgow.

Bednarik, R. G. 1979. The potential of rock patination analysis in Australian archaeology — part 1. *The Artefact* 4: 14–38.

Bednarik, R. G. 1986. Parietal finger markings in Europe and Australia. *Rock Art Research* 3: 30–61, 159–170.

Bednarik, R. G. 1990–91. Epistemology in palaeoart studies. *Origini* 15: 57–78.

Bednarik, R. G. 1992. The stuff legends in archaeology are made of: a reply to critics. *Cambridge Archaeological Journal* 2(2): 262–265.
Bednarik, R. G. 1993a. Refutability and taphonomy: touchstones of palaeoart studies. *Rock Art Research* 10: 11–13.
Bednarik, R. G. 1993b. Wall markings of the cave bear. *Studies in Speleology* 9: 51–70.
Bednarik, R. G. 1994. A taphonomy of palaeoart. *Antiquity* 68(258): 68–74. See also *http://mc2.vicnet.net.au/home/epistem/shared_files/antiquity94.pdf*
Bednarik, R. G. 1995. Metamorphology: in lieu of uniformitarianism. *Oxford Journal of Archaeology* 14(2): 117–122.
Bednarik, R. G. 1999a. Pleistocene Seafaring in the Mediterranean. *Anthropologie* 37: 275–282.
Bednarik, R. G. 1999b. Maritime navigation in the Lower and Middle Palaeolithic. *Comptes Rendus de l'Académie des Sciences Paris, Earth and Planetary Sciences* 328: 559–563.
Bednarik, R. G. 2002a. The dating of rock art: a critique. *Journal of Archaeological Science* 29(11): 1213–1233.
Bednarik, R. G. 2002b. Paläolithische Felskunst in Deutschland? *Archäologische Informationen* 25(1–2): 107–117.
Bednarik, R. G. 2003a. Seafaring in the Pleistocene. *Cambridge Archaeological Journal* 13(1): 41–66.
Bednarik, R. G. 2003b. A figurine from the African Acheulian. *Current Anthropology* 44(3): 405–413.
Bednarik, R. G. 2006. A unified theory for palaeoart studies. *Rock Art Research* 23: 85–88.
Bednarik, R. G. 2008. Children as Pleistocene artists. *Rock Art Research* 25: 173–182.
Bednarik, R. G. and M. Kuckenburg 1999. *Nale Tasih: Eine Floßfahrt in die Steinzeit.* Thorbecke, Stuttgart.
Bednarik, R. G. and M. Sreenathan 2012. Traces of the ancients: ethnographic vestiges of Pleistocene 'art'. *Rock Art Research* 29: 191-217.
Behrensmeyer, A. K. 1975. The taphonomy and paleoecology of Plio-Pleistocene vertebrate assemblages east of Lake Rudolf, Kenya. *Harvard University Museum and Comparative Zoology Bulletin* 146: 473–578.
Behrensmeyer, A. K. 1978. Taphonomic and ecologic information from bone weathering. *Paleobiology* 4: 150–162.
Binford, L. R. 1981. *Bones: ancient men and modern myths.* New York.
Brain, C. K. 1981. *The hunters or the hunted? An introduction to African cave taphonomy.* Chicago.
Cameron, D. W. 1993. The archaeology of Upper Palaeolithic art: aspects of uniformitarianism. *Rock Art Research* 10: 3–17.
Efremov, J. A. 1940. Taphonomy: a new branch of paleontology. *Pan American Geologist* 74(2): 81–93.

Gifford, D. P. 1981. Taphonomy and paleoecology: a critical review of archaeology's sister disciplines. In M. A. Schiffer (ed.), *Advances in archaeological method and theory*, Vol. 4, pp. 365–438. Academic Press, New York.

Guthrie, R. 2005. *The nature of Paleolithic art.* The University of Chicago Press, Chicago/London.

Hill, A. 1976. On carnivore and weathering damage to bone. *Current Anthropology* 17(2): 335–336.

Hill, A. 1979. Butchery and natural disarticulation: an investigatory technique. *American Antiquity* 44: 739–744.

Hiscock, P. 1985. The need for a taphonomic perspective in stone artefact analysis. *Queensland Archaeological Research* 2: 82–97.

Hiscock, P. 1990. A study in scarlet: taphonomy of inorganic artefacts. In S. Solomon, I. Davidson and D. Watson (eds), *Problem solving in taphonomy: archaeological and palaeontological studies from Europe, Africa and Oceania*, pp. 34–49. Tempus, Archaeology and Material Culture Studies in Anthropology, Vol. 2, University of Queensland, St. Lucia.

Solomon, S. 1990. What is this thing called taphonomy? In S. Solomon, I. Davidson and D. Watson (eds), *Problem solving in taphonomy: archaeological and palaeontological studies from Europe, Africa and Oceania*, pp. 25–33. Tempus, Archaeology and Material Culture Studies in Anthropology, Vol. 2, University of Queensland, St. Lucia

Sreenathan, M., V. R. Rao and R. G. Bednarik 2008. Palaeolithic cognitive inheritance in aesthetic behavior of the Jarawas of the Andaman Islands. *Anthropos* 103: 367–392.

Vishnyatsky, L. B. 1994. 'Running ahead of time' in the development of Palaeolithic industries. *Antiquity* 68: 134–140.

Werry, E. and B. Kazenwadel 1999. Garten Eden in der Sahara. *Bild der Wissenschaft* 4/1999: 18–23.

Ziegert, H. 2007. A new dawn for humanity: Lower Palaeolithic village life in Libya and Ethiopia. *Minerva* 18(4): 8–9.

Ziegert, H. 2010. *Adam kam aus Afrika — aber wie? Zur frühesten Geschichte der Menschheit.* University of Hamburg, Hamburg.

# 7

# A METAMORPHOLOGY OF ARCHAEOLOGY

Science expects exacting predictions for future observations about phenomena that can be measured. The regularities within these phenomena must be described as consistent patterns, explained by refutable theories cast in terms of causes. Metamorphology, as the science of how the perception of the individual archaeologist about what happened in the past relates to what *really* happened in the past, needs to analyse the epistemology of how archaeological data are collected, interpreted and disseminated. As a theoretical framework it is essentially predicated on the application of integral functions to all unknowns in archaeology. Metamorphology thus creates systematic uncertainties, but has the enormous benefit of being falsifiable, and of offering us unprecedented opportunities to test conventional, and otherwise untestable archaeological propositions. It needs to peruse each instance of inductive uniformitarianism and project it onto the canvas of the individual researcher's cognitive, perceptual, religious, political, ontological, academic and intellectual conditioning. It particularly needs to understand his or her limitations of knowledge concerning existing data, and consequences of language barriers and personal or cognitive biases. All of these factors act as very effective filters in how evidence may be perceived, interpreted and reported. The variables of the individual researcher's limitations should certainly not be immune to investigation; they are legitimate targets of research (see Chapter 6). And the number of possible systematic biases is incredible: confirmation bias, *déformation professionnelle*, selective perception, wishful thinking, wishful thinking bias, reactance, neglect of probability, Von Restorff effect, outcome bias, framing effect, bandwagon effect, expectation bias, congruence bias, attentional bias, clustering illusion (apophenia), conjunction fallacy, Hawthorne effect, observer-expectancy effect, primacy and recency effects, and most especially, selection bias. Several of these are amazingly common in archaeology and need to be given special attention.

## The collection of archaeological data

In the previous chapter we have seen that the basis of a sound epistemology of Pleistocene archaeology is the application of metamorphology. As its name indicates, this is the science of changes in the forms as which evidence seems to present itself, changes from the intrinsic to the subjectively observed. We can divide these changes broadly into physical and philosophically based types. The physical changes are those attributable to taphonomy — including decay, trampling, transport, mineralisation and many other processes affecting all kinds of material remains forming what is simplistically called the 'archaeological record'. The cognitive aspects of metamorphology concern cognisance of the way archaeological percepts are formed independent of sense-data (*sensu* Russell 1959: Ch. 1), as freestanding constructs of individual perceptions. Post-processualist archaeology, we noted in Chapter 2, fully accepts that all data are theory laden, and in the case of this discipline, their qualitative range and their quantifications are all determined arbitrarily, i.e. there is no falsifiable taxonomy to refer to. Thus the raw data of Pleistocene archaeology are collected in accordance with classification systems whose ultimate veracity is not accessible to external scrutiny (see Chapter 9). That does not render them necessarily false, but in using them it is essential that this be clearly appreciated. The artefact types we recognise, tabulate and record are generally invented constructs ('archaeofacts'), as are many other entities recorded as supposedly empirical phenomena, ranging from perceived relations to landscape to interpretations of intent.

Even the apparently 'more scientific' aspects of such practices can and need to be questioned. For instance one of archaeology's most important practices is the production of perceived sediment stratigraphies from excavated profiles. Without it, there can be little archaeological discourse. And yet, this could be demonstrated to be a very subjective procedure. Suppose one were to conduct a 'blind test', locate a complex stratigraphic section and ask, say, ten archaeologists to independently draw the same deposit's stratigraphy. Almost certainly this would yield ten different drawings, 'ten different stratigraphies'. The reasons for this are diverse: proper sedimentary analyses are not conducted during most excavations; samples might be taken but are usually processed much later in a laboratory. Most archaeologists have inadequate understanding of sediments (I say that as one who has operated a sedimentary laboratory since the 1970s), and therefore these section drawings are generally the result of simple eyeballing. They are essentially individual artistic and unique works, which means that their derivation is not a repeatable process, i.e. it is untestable. In practice, only one such section drawing is usually secured of a particular stratigraphy, and then the section itself is most often destroyed by continuing excavation or by the elements. Therefore no procedure of refutation is feasible; we have to take the recorded stratigraphy purely on trust; and yet it contains the only surviving evidence of it.

It is not suggested that this practice should be revised; rather, it just needs to be illuminated here what has to be understood in the context of metamorphology. We need to appreciate that the section drawings of sediments, although often of extraordinarily high analytical quality, are not necessarily statements of 'facts', but are *interpretations* whose veracity depends solely on the competence of the recorder. The kinds of sediments archaeologists tend to excavate are complex deposits that are not readily understood, and yet the veracity of archaeological claims hinges largely on their correct interpretation. Similarly, the recordings made of rock art are not true transcriptions, but abstractions and interpretations of what is on the rock. This has been legally recognised with a High Court ruling in Austria that rock art recordings are subjective *works of art* and therefore subject to copyright law. Just as our taxonomies of artefacts derive from 'egofacts' (Consens 2006), a variety of further dimensions the Pleistocene archaeologist may perceive as facts are also self-confirming phenomenon categories. In Searle's (1995) terminology, they are 'institutional facts'.

This is readily illustrated with rock art, which is traditionally approached from two alternative but complementary directions by archaeologists: they either seek to determine its meaning (especially what it depicts) or invent a taxonomy of its motifs, often to use as the basis of creating purported sequences of traditions. Although both these approaches lack epistemological justification, they constitute essentially what is generally considered to be an 'archaeology of rock art' (Chippindale 2001; Chippindale and Taçon 1998). Archaeologists are hardly in a better position than other cultural aliens to determine what is depicted in rock art. They derive these interpretations by the same means as everyone else: through autosuggestion, the psychological process by which the individual induces self-acceptance of an opinion or belief. In the absence of any sound ethnographic input, taxonomic entities or systems of rock art are autogenous, etic constructs of untutored observers, irrespective of the training of such people. To place this issue into perspective it needs to be considered that the belief of archaeologists to have some unexplained access to the meanings and intents of ancient peoples can be easily tested by comparing their practice to that of art critics of contemporary artists. Modern-day artists notice regularly that the self-appointed 'experts', the art critics, fail in comprehending the meaning and intent expressed in modern art works of artists of their own cultures (Tony Convey, pers. comm. June 2010). What hope is there that (self-appointed) archaeologists can do this with the works of cultures they have no connection with? They are probably not even aware that there are significant differences between the brains of literate and non-literate people (Helvenston 2013), attributable to the ontological modification of both the chemistry and structure of the brain through affecting the flow of neurotransmitters and hormones (Smail 2007) and the quantity of grey matter (Maguire et al. 2000; Draganski et al. 2004; Bednarik 2013).

The subjective constructs derived from invented taxonomies of rock art are routinely used as a basis of considerations of style, and in attempts of defining cultural traditions and their sequences (Figure 50). This approach is the basis of 'archaeological dating' of rock art, which relates these etic definitions

*Figure 50. Typical complex rock art panel, a favourite subject of archaeologists in inventing styles, traditions and cultures.*

to cultures and their sequences that have been perceived through excavating occupation layers. The epistemology of this approach contains numerous logical faults. To begin with, the 'cultures' the Pleistocene archaeologist names and defines are not reflections of reality; they are inventions, based on modern opinions of self-appointed experts of the human past about what constitutes cultural traditions of these ancient times. There is no testable or refutable evidence that these 'cultures' marked any ethnic, political, social or linguistic groups, tribes or nations. They are identified largely on the basis of minor differences in the stone tools they used. Obviously tools do not define cultures, their characteristics merely reflect ideas and practical conventions. Such ideas or 'memes' may travel or they may be re-invented independently elsewhere. Here we meet again this fundamental trait of orthodox Pleistocene archaeology: the notion that movement of qualitative indices of material finds proves movement of people, and here it is stone tools that are the fetishes representing people. Art or symbolic behaviour, by contrast, is far more culture specific, but the discipline has historically chosen to base its taxonomy on stone tools, i.e. on technology rather than culture.

So in Ice Age archaeology, cultures have not been defined correctly, because instead of determining cultural taxonomies we have established what appear to be technological taxonomies. To then force cultural variables such as palaeoart into these non-cultural pigeonholes is absurd. It would have been more logical to try establishing cultural parameters and sequences first, and to then fit the technologies into this framework. But even this is usually not possible, and it is again illustrated by the ways archaeology approaches rock art — which are in fact amazingly 'un-archaeological'. Typically, major sites comprise complex superimposition sequences spanning various cultures, because widely separated societies contributed to the same panel. Some of the artists may even have reacted to pre-existing art that no longer survives. Yet the archaeologist, using a low-resolution approach to the 'study' of this cumulative corpus, tends to treat it as a single entity, even as a representative sample. Unable to determine the ages of these chronological components, the archaeologist might invent styles to separate them. But where physical superimposition of motifs is largely lacking, as is most often the case, these 'stylistic constructs' are again merely freestanding formulations generated by autosuggestion. So in contrast to an excavation, which offers a layering of successive traditions in three-dimensional space (with some caveats), the successive traditions of a major rock art site occur in two-dimensional space; they have no archaeological time depth. In many cases archaeologists then assume that only one tradition has contributed, and they call this a style. There are numerous examples of such 'styles' that are in fact the precipitates of several traditions (e.g. the 'Panaramitee style' in Australia; Bednarik 2010) (Figure 51). This practice of lumping together all components of a site is logically identical to excavating layers of the Palaeolithic, Neolithic and Bronze Age and then lumping together all artefacts to define them as belonging to a single culture. No archaeologist would do this with excavated remains (at

*Figure 51. An invented style of Australian petroglyphs, the 'Panaramitee style'. It has no objective existence, it comprises elements of different periods arbitrarily lumped together.*

least not intentionally), but with rock art it is a frequent practice, unless there is some really blatant contrary evidence available.

The example of rock art to illustrate how an epistemological analysis shows that the fundamental assumptions about sites may be completely wrong can be extrapolated to many other practices or areas of Pleistocene archaeology. Indeed, there are many scenarios in archaeology where the correct interpretation of a variable may be the exact opposite of what the researcher might be inclined to think. For instance there is the tendency, already touched upon in the context of rock art, to make implicit assumptions about different evidence types being related because they occur at the same site. Most sites excavated by Pleistocene archaeologists are 'favoured localities' in the landscape, for obvious reasons: it is where one tends to find things. They may be caves, rockshelters, desert springs and so forth. Many such sites have been lived in by tens of thousands of individuals (over hundreds of millennia). Therefore the probability that any two unconnected forms of evidence (e.g. rock art and a hearth) date from the same visit is infinitesimally small. Conversely, the probability of any two forms of anthropic evidence being of the same visit is millions of times greater at a featureless site in a huge featureless plain (Bednarik 1989).

Clearly, then, there are systematic and hence assessable cognitive biases in the way data are collected in the field — biases that relate to the collecting researcher: to her/his methodology, funding limitations, time constraints, limitations of knowledge and expertise, and other such variables. As any practitioner will readily admit, the quality of archaeological fieldwork varies enormously, and such factors are to some degree measurable. But how does one quantify factors such as individual knowledge with any semblance of objectivity? It is obviously possible, with some diligent research, to form subjective views of a researcher's competence, but this would at best provide a very coarse tool of analysis, and be misleading at worst. It is the task of metamorphology to develop such principles in testable, repeatable and transparent formats; *repetitio est mater studiorum* (repetition is the mother of science).

## A review of archaeological interpretation

Unless we can understand how data were collected there is no metamorphological depth in the presentation of observations. But not only are data *collected*, they are also *stored*, *interpreted* and *disseminated*. Each of these steps is subject to a whole range of biases. For instance, archaeologists may assume that charcoal contained in a sediment layer indicates the age of the layer, relying on an associative hypothesis. In an epistemologically sound system all components of a sediment layer are of different 'ages', and there is no guarantee that charcoal can date the event of the sediment's final deposition. Numerous common processes can distort the relationship. Secondly, most archaeologists mistakenly believe that charcoal is the age indicated by its content of carbon isotope $^{14}C$. This is false for several reasons, among them the fact that this index does not even relate to the event of carbonisation of the wood, but to a process of assimilation in a living tree. Or in epistemological language: we have a supervenience condition between two propositions, both of which are factually false. Yet archaeologists habitually speak of the 'radiocarbon age of a sediment', which is a nonsensical concept.

Or to use a differently framed example to illustrate the endemic issue, consider the idea of micro-wear analysis. It may be very useful and seems to exhibit the trappings of a scientific method, but it is *not* scientific; it does not satisfy the requirement for falsification. The observation that microscopic wear on an archaeological specimen resembles experimental wear on a replicative one does not prove that two similar processes pertain. Again, the dependency relation is one of supervenience: there could be a difference in one set of properties without there being one in the second. In other words, *modus ponens* is not valid. The proposition that the traces on the ancient specimen were made by the same process as those on the replication artefact is not refutable. It may well be true, it quite probably *is true*, but that is not the issue. The issue is one of falsifiability, which in all of archaeology simply does not apply — not without importing methods of a hard science.

We have considered some variables affecting the methods of collecting the field data, using a small number of examples to illustrate falsities in acquisition strategies. But then this gathered material is also stored, interpreted and disseminated. At each of these stages there are subtle but systematic sources of contamination, error or bias. For instance storage and curation of Pleistocene osseous finds leads to several types of contamination affecting their susceptibility to analytical work. Excavation and storage destroys most DNA (Pruvost et al. 2007), while preservation treatment tends to eliminate susceptibility to dating analyses. The priorities in what materials to save and store and what to discard are of considerable consequences, and they are again subjected to cognitive biases, which may amplify those of the collector. Which finds and materials are retained is to a large extent determined by the excavation's professed purpose, perhaps the orientation of those conducting, funding or approving it, and most certainly by limitations of knowledge on their part. For instance the most important archaeological component of any sediment at the foot of a petroglyph panel are the stone hammers used in the production of the petroglyphs (Figure 52), because their stratigraphical position is likely to tell us roughly at what time petroglyphs were made at the site (Bednarik 1998). Thousands of such deposits have been excavated, usually in the hope of exposing rock art below ground in order to determine mere *minimum* dates for the art. This has succeeded in very few cases globally (Daleau 1896; Lalanne and Breuil 1911; Capitan et al. 1912; Lemozi 1920; Hale and Tindale 1930; David 1934; Passemard 1944; Ampoulange and Pintaud 1955; Mulvaney 1969: 176; de Saint Mathurin 1975; Anati 1976: 34, 41; Thackeray et

*Figure 52. Stone hammers used in the production on petroglyphs at a site in the Pilbara, Australia.*

al. 1981; Rosenfeld et al. 1981; Cannon and Ricks 1986; Steinbring et al. 1987; Bednarik 1989; Crivelli et al. 1996; Roberts et al. 1998; Pessis 1999; Bednarik et al. 2005). But in the process of this mostly futile pursuit, the chance of securing *actual* ages of petroglyphs via stratified hammerstones was literally destroyed. An example are the over one hundred excavations in the Côa valley of Portugal, where not a single submerged petroglyph was found, yet the stone tools used in the production of the rock art were discarded because the archaeologists had not been trained in recognising them (Swartz 1997; Bednarik 2004). This is a classical example of how inadequate knowledge destroyed the research potential of sites, and there are many other examples like this. For instance, nearly all the world's Lower Palaeolithic finds of wood and resin, thousands of them, have been reported from just two countries, Germany and Israel. It would be foolhardy to suggest that such remains could only survive in these two regions; it is much more likely that archaeologists there are more adept in recognising or securing fragile organic materials from excavations.

Biases in the eventual interpretation of the data are even more decisive, particularly as they are likely to change as the discipline's ideology vacillates historically. Pleistocene archaeology is entirely at the mercy of the historical sequence in which key discoveries are made — those that guide the dominant paradigms. In contrast to the systems of data gathering in most other disciplines, there can be little design in the data acquisition strategies of Pleistocene archaeology. Most key finds are made fortuitously, yet they may decide how other aspects are interpreted. For instance when the period's first rock art was reported, from Altamira in Spain, it was completely rejected as we have seen in Chapter 4. Its sophistication was considered entirely incompatible with the perceived primitiveness of Upper Palaeolithic people, as deduced from their earlier found tools. Yet it is obvious that if Palaeolithic cave art had been reported and accepted first, it would have been the tools that would have been rejected as being contemporary, because *they* would have been regarded as incompatible with the sophistication of the art. In either case the perceptions and expectations of scholars will be significantly distorted, yet the order in which discoveries are made and accepted is largely accidental. Similarly, their acceptance depends on perfectly subjective factors at any time, and that certainly has been the case since the 19th century and *has continued right to the present.* Today many Pleistocene archaeologists reject the idea of symbolism and palaeoart prior to 35,000 years ago, because they believe they already know what the cognitive levels of earlier humans were. Nothing has changed in the epistemology of Pleistocene archaeology, which suggests that we need to expect to see blunders as monumental as those of the past still occurring today.

The only thing systematic in this interdependence of sequence of discoveries, currency of paradigms and political currents of various types is that there will always be systematic distortions in the positing and acceptance of interpretations. These patterns can only change in kaleidoscopic modes as new evidence and new theories become available. Moreover, a powerful driving factor of archaeological theories about Pleistocene hominins is public perception. Rather than influencing it, the discipline ingratiates itself with the public, on whose support it depends entirely, by constantly re-inventing itself according to social currents of society. The Neanderthal flower children of the 1960s became the victims of competition in the cynical 1990s, but today, with the failure of the market-driven economies and the spectre of man-made global climate change, their fortunes look much better again. In a few years their correct name, *Homo sapiens neanderthalensis*, will once again be reinstated by a discipline that changes its spots every time society modifies its views and priorities. One of the most disappointing aspects of archaeology is that it does not affect public perceptions; instead it merely adopts, amplifies and justifies them. It thus lacks authority and authenticity, and the only certainty about archaeological interpretations is that they will always change, previous ones will be recycled, and new versions will be added from time to time. Clearly, this is not a scientific form of discourse and development. It is entirely dominated by non-scientific currents, and it is metamorphology's monumental task to unravel these intricate currents, to explain them, their interplay, and their effects.

## Biases of dissemination

The distortions resulting from the modes of collecting, storing and interpreting the materials Pleistocene archaeology is interested in are so great

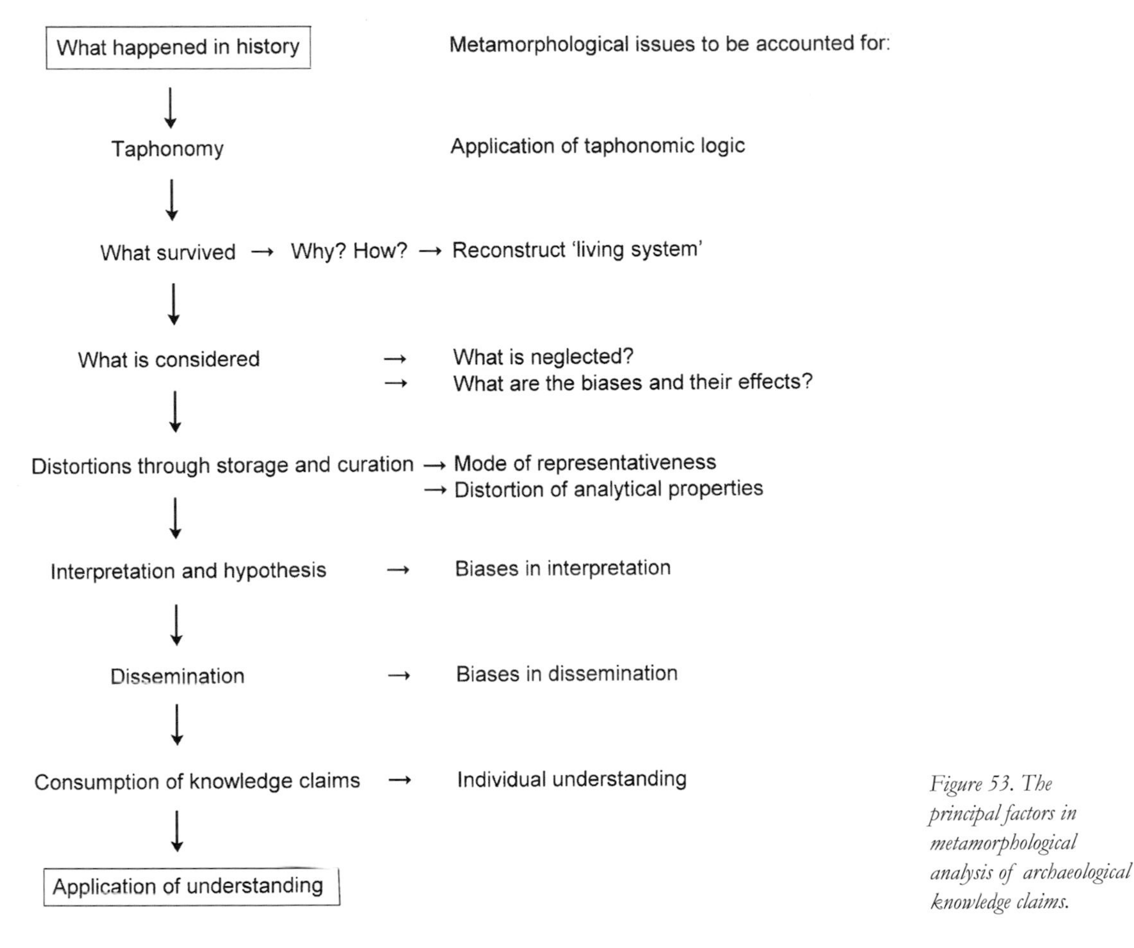


*Figure 53. The principal factors in metamorphological analysis of archaeological knowledge claims.*

that the mission of metamorphology is indeed a huge endeavour (Figure 53). In addition we still need to consider the distortions that inevitably occur in the *dissemination* of these data. The principal formal venues of dissemination are archaeological journals, which operate much in the same way as those of scientific disciplines. Papers about finds and interpretations are submitted, and are refereed by specialists in the particular field the author addresses. This system of peer review may work well in other fields, but in a non-scientific discipline it can become an agent of amplification of biases. The analyst then needs to understand the possible sources and principles of systematic processes. To begin with, an editor will select referees on the basis of her/ his own experiences, and will be guided by several subjective factors. Having been an archaeology editor since 1965, and having edited several journals and dozens of monographs, I have some relevant insight. As in any field, intellectual cliques have formed in specific subject areas of Pleistocene archaeology, whose members may have developed narrow foci and exclusive views. If the submitted paper complies with their collective model, they are likely to recommend its publication — often after requesting that their own work be cited in it. However, if the paper strays too far from the axioms of an influential paradigm, it is almost impossible to secure its publication, irrespective of its veracity or merits. The referees are likely to subscribe to a broadly based consensus model, and if there is no profoundly partial evidence in favour of the paper's contentions, they will be unwilling to forego the rewards that go with being the 'gatekeepers' of what is acceptable (a position they acquired through their being regarded 'reliable'). By the same token it is also true that the paper needs to present something new and different in order to be worthy of publication. Therefore an author's best strategy to get work into print is to endeavour offering some new ideas or data, but without upsetting the established order of the sub-discipline being addressed. The kind of paper finding the most ready acceptance presents new material that, in effect, confirms the existing dogma, and thus reinforces the positions of the 'gatekeepers'.

This role of disciplinary inertia is reinforced by other factors, among them the established hierarchy of scholarly serials. The world's archaeology journals are ranked according to perceived influence, prestige and citation factor, which are essentially self-fulfilling perceptions. It matters greatly *where* the new work is published, but the greater the authority of a journal, the greater the reluctance of its editor and its referees to relinquish control of the dogma. They then have no choice but to reject a heretical paper that challenges the very foundations of the discipline's beliefs, and they may thus be inclined to reject even mildly iconoclastic work. While the hard sciences are required to submit to falsification, a non-falsifiable pursuit such as archaeology is not obliged to allow challenges. Therefore works of major heretical impact are extremely rare in this field, and should only be contemplated by strong individuals of immense dedication and intellectual resources (see Chapter 4). Alternatively they need to be submitted to minor journals, which of course renders them difficult to access and rather ineffectual.

Finally, in reviewing the dynamics of the dissemination of archaeological knowledge claims we arrive at the consumer of this information. If there is no collective subconscious to which all practitioners are somehow connected, and no uniform world archaeology or standard knowledge, as proposed here, it is inevitable that the reception of the conveyed information will differ significantly among recipients. The way it will be received, considered, processed and, especially, *applied* by the individual will vary among archaeologists, paradoxically in spite of the need to observe dogmas. This is particularly so as there are as many different archaeologies as there are practising archaeologists (see Chapter 2). These variations will be determined by individual knowledge, ability to reflect (e.g. upon criticisms), preoccupations or priorities, aspects of personal disposition (e.g. temperament), time constraints, and most especially by the prior conditioning of the practitioner concerned. All of these many factors need to be accounted for in a rigorous metamorphological understanding of how the individual receives and uses archaeological information to make sense within his or her reality construct.

None of this is intended as a criticism; it is mentioned here only in the context of the importance of these issues to the metamorphologist. She or he needs to understand the processes and pitfalls of dissemination: how there are systematic distortions in what has been found, analysed, considered and written about, because of selective reporting. 'Unpopular finds' will not be reported in mainstream international journals, irrespective of how important they are (the Berekhat Ram proto-figurine is an example; Goren-Inbar 1986). Consequently very few scholars will even be aware of them. Favoured paradigms will dominate the orthodox discourse, evidence supporting them will be over-emphasised, and evidence opposing them tends to be censured by the referee system. An example of the latter is the Professor Reiner Protsch affair in Germany (see Chapter 5), which was *not* hushed up because it discredited one academic, but because it significantly damaged the 'African Eve' hypothesis, and thereby affected thousands of academics. Thus these kinds of faddish theories may survive through practices of misinformation, and the hegemonic routines augmenting them need to be fully understood by the epistemologist of Pleistocene archaeology. As we have seen in Chapter 2, we have an endless variety of archaeologies around the world (Figure 54), but by far the most prestigious and authoritative grouping is the Anglo-American school. It alone determines archaeological dogmas these days, yet it is among the most 'inbred' traditions. In such a near-monopolistic environment, sectarian interests moderate the wider dissemination of knowledge and pluralist notions are discouraged. This renders the discipline quite susceptible

*Figure 54. Animated archaeology.*

to the amplification of mistakes, which are not only such a dominant factor in its past but remain so to the very present. There is no reason, from an epistemological perspective, why today's dominant models of Pleistocene archaeology could not be as spectacularly false as those we have visited in Chapters 4 and 5. The discipline's epistemology remains as defective as it was from the beginning.

## The blowtorch of logic

Metamorphology can address this issue most effectively — which may well be why mainstream archaeologists are shunning it. Just like taphonomic logic, it tends to apply the blowtorch of logic to the fantasies of archaeologists. In my work as archaeological editor I have seen thousands of papers that were not published because I had to reject them. So far this chapter has been an excursion into theory, and much of what has been said is rather abstract and perhaps not best suited to readily elucidate the issues raised here. A more effective way to communicate concerns with the average levels of archaeological understanding of archaeologists is to illustrate them by presenting a specific example and commenting on it analytically. I emphasise that the following example was selected entirely randomly; it is a passage from an academic paper I had to check only in the same week as writing this text. This is not a particularly negative example (I could provide hundreds of similar or worse ones), it is from a paper by a university lecturer who holds a doctorate and professorship in archaeology. It describes a petroglyph site (Figure 55) and this text is from a chapter entitled 'Analysis and interpretive reflections':

*Figure 55. Section of the petroglyph sites described here.*

> Rock is immovable and permanent; for this reason men of all ages have used it to make graphic the ideas of their time, selecting flat surfaces of rock, often smoothing them to adapt the designs to the spaces available. Technical evidence indicates that the executors worked the designs directly on the granite rock, using soft percussion, pricking the segments of rock impacted; for rubbing, they might have used deer horn and hard wood. What is most certain is that the petroglyphs were created, leaving messages in their motifs of cupules and visual figurations for us to study today.

Rock is neither immovable nor permanent, in fact many quartz grains have undergone repeated phases as sand and sandstone, and the archaeologists who make a living removing rocks with petroglyphs would understandably disagree with the first statement. The next part of the first sentence is an unsubstantiated platitude: in nearly all cases we do not know the reason(s) why rock art was created, and seen in the context of taphonomic logic, the statement is nonsensical. Next we have the assumption that rocks were 'smoothed' before markings were made. This is an extremely rare practice and has not been observed in the region concerned.

Moving to the second sentence, we find the absurd assumption that petroglyphs on very hard rock (granite, apparently) were made by 'soft percussion' and 'pricking' (whatever these terms are supposed to mean), and by abrasion with antler and wood. This shows that the author lacks understanding of the relevant materials and technology. The remaining sentence implies the attachment of messages, which the archaeologist can 'read', i.e. it is self-delusional. So we move on to the immediately following paragraph:

> In the upper stone, the arrangement of the cupules is circular, becoming more concentrated from outside to inside; that is, toward the interior of the carved space; the arrangement of the holes gives a sense of the parallelism imagined from ancient times in connection with the cosmic order and the existence of man, according to which intricate relations exist between the course of heavenly bodies and the thought of man. The changeless appearance of the engravings in the rocks would have converted the site into a sacred place. The proportions of relational symmetry achieved through the sculptural realism of a circular nature turn the group of cupules into an analogy representing the exterior space, where man would consider his reality to have been transmuted.

We see how after a half sentence of description, the author abruptly delves into interpretation, which immediately descends into meaningless clichés. Here, the notion of 'interpretive reflections' becomes a euphemism for unfettered imagination and fantasy. Every single statement and phrase is offered without evidence, supporting reasoning, justification, or deductive warrant. Indeed, the next paragraph consecrates the site without hesitation:

> The cupule is indifferent to the type of rock in terms of converting it into an altar of significant social value; the cult or ritual does not celebrate the rock itself; its significance arises from the encounter and where the ritual takes place. The second (lower) engraved stone displays a concentration of incised sculpted motifs with more variety, giving form to linear-geometric motifs that expand our information about the relation between the figurations and ideology. The three visible incisions form a grid in the space; its distribution in four parts would symbolise the spatial balance intuited for the cosmos, which the eyes of man distinguished in the succession of night to day.

At this point we need to be reminded that the site has not been described, we have little idea of what it comprises, of its morphology, topography or setting — or indeed any scientifically relevant information that might help us form an independent view of the site and its rock art; we are simply presented with a fait accompli identification as a ritual site of some cosmic properties. These interpretational inanities are presented as legitimate archaeological information, and they are justified in the next paragraph:

> Thus, we insist that actively analysing the iconography is more important than simply describing the motifs. Among the constituent conditions of the acquisition of knowledge regarding the concrete realities of the world and universe is the perspective of our own view of the world, and the time in which we live.

These statements are most apposite in understanding the epistemology of archaeology. There is, first, the author's autosuggestive belief that his notions amount to an 'analysis'. The word *analysis*, however, defines the separation of an entity into its constituent parts, whereas the various musings of the author seem to amount to a *synthesis* of subjective observations (vibes), personal beliefs and very vague ideological notions about people of the past (always bearing in mind that the author has no idea how old the rock art is, and therefore to which period or people it could possibly refer). The second sentence provides an interesting illumination of the author's professed perception of epistemology. He seems to espouse a laudable relativist "view of the world", in which our perspectives are mere "constituent conditions of the acquisition of knowledge". But if he is guided by this understanding that we, the present people, do not adequately understand the world we exist in, how can he possibly take it upon himself to 'explain' the world or the motivations of past civilisations? Especially through a rock art he has no scientific grasp of, that he cannot attribute to any civilisation, and that he cannot place into a framework of testing by taphonomic logic.

This brings us back to the fundamental shortcoming of archaeology. If we accept that as a species we are not in a cognitive state to fully comprehend our own reality (although the neurosciences can and will solve this), it would appear somewhat premature to blindly probe into the realities of others that lived long before us, whose worlds we cannot begin to comprehend. Surely if we sought to explore the processes that led to the particular constructs of reality we subscribe to today, we would need to fathom their origins, which would presumably involve a rigorous exploration of the cognitive development of societies before ours. But how can we entrust such a delicate and demanding quest to a discipline that has so far only managed to generate a cacophony of mythologies about the distant past that are in reality often only reflections of *contingent beliefs and contemporary social constructs* (which we also understand inadequately)? It would be precipitate to base an inquiry into how hominins acquired their various constructs of reality on the models and views developed by such an undisciplined discipline as Pleistocene archaeology. It is here that metamorphology becomes indispensable in correcting it, and neuroscience in solving the issue (Bednarik 2011, 2012).

Metamorphology, like taphonomic logic, not only cuts through the bombastic verbiage that marks so much of archaeology, it does this with the greatest ease and effectiveness. Consider the example of the claim that the locations of Palaeolithic cave art prove that rituals took place in these caves, which we visited in the preceding chapter. In analysing this claim, metamorphology begins with observations of taphonomic logic: what is the 'crucial common denominator' of the phenomenon category (Bednarik 1990-91, 1994)? It finds that the CCD of 'cave art' is most probably not location, but almost certainly selective preservation derived from location. It then moves on to analyse how, beginning with a sophism, an entire mythology has been developed around it (e.g. that it is the oldest palaeoart). It then dismantles every part of this chain of probably false interpretations by showing that alternative explanations are either just as plausible, or, in most cases, even more so. Thus metamorphology transforms the dogmatic 'certainties' of Pleistocene archaeology into ambiguities, creating the systematic uncertainties science needs before it can consider scenarios of probabilities. It creates reliable knowledge by untangling the historically ossified 'received knowledge', separating it into its constituent parts, determining where these came from, how they were developed, who championed them, how reliable their basis really is, and other such aspects that, collectively, form the *epistemology of Pleistocene archaeology*.

And this is precisely where our task begins.

**Return to Eve**

At this point it is requisite to return to one of archaeology's most powerful dogmas of recent times, which we visited in Chapter 5: the 'African Eve'

*Figure 56. When the Moderns met the Robusts.*

model (Figure 56). In analysing it, the epistemologist could begin by asking: what are its constituent parts, where do they originate, how was this hypothesis developed and by whom? We have seen that this theory lacks supporting archaeological evidence: there is nothing to suggest that the Eurasian Upper Palaeolithic technologies or palaeoart traditions were introduced from Africa. Rather, the evidence suggests that they were locally developed twenty to thirty millennia before the Middle Stone Age of northern Africa gave way to the Later Stone Age (Bednarik 2008a, 2008b, 2011). I have suggested that the search for modern human origins is itself a misguided quest, because modernity is a function of culture and cognition, not of facial features or minor skeletal developments. Then there are the revelations of recent years that nearly all human remains of Europe which the African Eve advocates had cited in the creation of their theories had been misdated, or their datings were in fact fake. Ultimately the model relies mainly on genetic differences among populations *of different eras*, and the principal support, after discarding all misconceptions and erroneous claims, derives from the notion that the distribution of modern genes reflects the mass-movements of people out of Africa. Minor support is also sought in the evidence that robust *Homo sapiens* people called 'Neanderthals' show slight differences in their DNA, relative to contemporary people.

Here we are not concerned with refuting this model, but only with how it came into being, how it established itself as dogma in much of the world, how it might illuminate the epistemology and promotion of dominant paradigms in general. It developed from a preceding, similar idea (Bräuer 1984) that was substantially based on the 'hoaxes' of Professor Protsch (Schulz 2004). By the late 1980s, British and American researchers had formulated the 'Eve' model, especially after a team at the University of California at Berkeley had subjected

136 mitochondrial DNA samples to a computer program designed by Alan Templeton, attempting to construct a family tree for 'modern humans'. They reported that we must all descend from one common mother that lived about 200,000 years ago. Dr Templeton then pointed out that the same data could have generated $10^{267}$ alternative and equally credible family trees (which is very much more than the number of elementary particles of the entire universe, about $10^{70}$!), and the announcements were thus attributable to a computer bungle. Many other objections have been voiced, among them the apparent morphological continuities in European and especially Asian hominin populations, and Alan Mann's earlier finding that tooth enamel cellular traits showed a close link between 'Neanderthals' and present Europeans, which both differ from those of Africans (Weiss and Mann 1978). This has recently been confirmed by genetics (Green et al. 2010).

Instead of abandoning their hypothesis, the Eve advocates tinkered with its details and continued promoting it so aggressively, through a few dominant journals, that it won rapid public approval. An important factor in the popularity of this model was the perception that it underpinned the idea of a single humanity, whose individuals are all ultimately related. But in fostering this feel-good notion of togetherness its academic backers overlooked two potential ideological objections: their tale of the rise of our ancestors who exterminated or out-competed all other humans on the planet involved a sinister side also (Bednarik and Kuckenburg 1999); and academic spin may foster academic careers, but it is detrimental to scientific veracity. At best, the claimed glorious triumph of our forebears would have come at a terrible cost to other humans; at worst it endorses fierce competition to the point of extinction and even becomes a rationalisation of genocide.

The implicit argument is that the Robusts are no longer around to object or suffer any ill effects, whereas the doctrine of universal kinship could bring today's nations closer. Academic partiality guided by political rationalisation is common in archaeology, but it should not guide science. In the first few years of the current century, the tide began to turn against Eve, despite strenuous endeavours by her advocates to maintain the momentum of their crusade. Having probably realised that the cultural and technological evidence would yield no support for their cause, and that even their ascendancy in the palaeoanthropological arena was under serious threat (particularly from the troublesome intermediate morphologies of so-called 'hybrids'), they focused increasingly on the new techniques of genetics. The thrust of the replacement scholars' tenet became to emphasise, as much as possible, the differences between hominins regarded as robust and those regarded as being Eve's progeny. Bones of 'Neanderthals' were analysed, and minor differences in very fragmentary (and contaminated) DNA sequences were hailed as evidence that they must have been a different species. Moreover, the present distribution of DNA markers, it was claimed, indicates that today's humanity spread exclusively and relatively recently from Africa.

Again, we will ignore here why these claims are false (but see e.g. Vigilant et al. 1991; Barinaga 1992; Ayala 1996; Templeton 1996, 2002, 2005; Kidd et al. 1996; Brookfield 1997; Harpending et al. 1998; Pennisi 1999; Strauss 1999; Adcock et al. 2001; Fedele et al. 2002; Gutierrez et al. 2002; Hardy et al. 2005; Garrigan et al. 2005; Fedele and Giaccio 2007; Green et al. 2010), and focus on how they might have come about. The notion of tribes wandering through uninhabited landscapes is particularly prominent, forming the demographic canvas facilitating the externalisation of this gene fetishism. Colonisers of empty expanses of land are seen as the vessels of genes, and genes come to represent populations. Both assumptions are solipsisms: given enough time, genes can travel to the ends of the world without any mass movement of adequate numbers of people to replace resident populations (through successive generational mating site distances). And the idea of empty spaces permitting these Exodus-like migrations is as absurd as would be a belief that the Biblical Exodus did not result in the displacement of other tribes. Just as all ethnographically known hunter-forager-fisher peoples have occupied virtually all habitable regions of their world, the robust *Homo sapiens* people of Eurasia and Africa (and eventually Australia) had settled any part of these continents that was even remotely fit to live in, given their technologies. For instance the 'Neanderthals' have lived in the far north of Europe, even inside the Arctic Circle, where temperatures in the last Ice Age would have been well below today's -40°C at times (Norrman 1997; Pavlov et al. 2001; Schulz 2002). This shows, firstly, that these supposedly primitive people must have been technologically for more advanced than most archaeologists give them credit for; and secondly, that they were prepared to accept such extremely harsh conditions. We can therefore safely assume that the more liveable parts of the continent were all occupied before the mythical invaders (Bednarik 2008a) from the tropics could have arrived. The residents were infinitely better adapted, both physically (having evolved in European conditions for several hundred millennia; Caldwell 2008) and technologically. Indeed, their culture and technology was clearly superior to that of northern African or Levantine populations between 50,000 and 30,000 years ago. Moreover, they were physically far more powerful (as we know from their skeletal muscle attachments) and had much more robust skulls and skeletons than the fragile Eve descendants are said to have introduced. The very idea that these well-adapted, technically superior, extensive robust populations that covered most of Europe, with their vastly superior arsenal of adaptive alleles, would have allowed bands of unclothed tropical invaders to push into their bitterly cold territory and eventually wipe them out looks therefore rather preposterous. The robust residents would have simply swamped the newcomers with their genes, with their better adaptations, their superior cognition (as evidenced by their use of palaeoart), their resident status (any military strategist knows how hard it is to displace a resident population in armed conflict) and with their much greater numbers, physical strength and robusticity. The only sensible argument here is that of new diseases, but this works of course both ways.

Since we can safely assume that all suitable parts of the Old World were occupied during the Late Pleistocene, how did this false idea of largely unpopulated regions arise? It is clear from the perusal of the literature of the replacement advocates that they tend to see the distribution map of hominin finds from the period as somehow reflecting actual populations: one is a fetish for the other. In the context of the earlier initial peopling one writer sees Asia becoming populated by a "chain" of 37 bands totalling 1110 people, expanding from Sinai to the Bay of Bengal, where they split into two "chains", one of 61 bands and leading to Lantian in China, the other ending at Modjokerto in Java and comprising another 33 bands (Webb 2006: 20). This colonisation scenario involved a total of 2820 people, and like all others has no credible basis of any kind. It is pure fiction, as are most scenarios invented by the discipline.

Again there is an archaeological fetish involved: fossils that subliminally represent people and populations (human fossils in the Levant, in China and Java). This illustrates once more how taphonomically illiterate most Pleistocene archaeologists are. Because they fail to comprehend the implications of taphonomic logic, they form entirely false constructs of the past. All variables relating to hominin remains, like those of all other animals, have undergone massive taphonomic distortions. On the whole, it is amazing that any have survived at all, and this is always attributable to fluke preservation conditions. Such materials only occur on very rare occasions, mostly in sheltered high-pH sediments that have not been subjected to such factors as frost action. Moreover, few of those that have survived have actually been recovered so far. To then derive from the map of their distribution (or of preserved occupation sites, for that matter) demographic deductions about former populations is illogical. What such a map *does* reflect is the distribution of where the best preservation conditions applied, *and* where researchers have so far looked. Again we see how expeditiously metamorphological examination detects fallacies in interpretation.

A sensible null-hypothesis would be to assume that, 45,000 years ago, all environments of four continents (plus at least twenty islands) permitting human colonisation were as densely occupied by hominins as their carrying capacities permitted. Hence there were contiguous populations from southern Africa to Japan. These already possessed regional characteristics that had evolved over many hundreds of millennia in response to specific environments, diets and lifestyles. Reticulate introgression, genetic drift and episodic genetic isolation had occurred throughout, as Weidenreich's multiregional hypothesis had long suggested (Figure 57). Most importantly, the movement of genes is much more plausibly explained by allele drift based on generational mating site distance rather than by mass migration (Harpending et al. 1998). A mating site distance of merely 100 km per generation is most reasonable for such highly mobile populations, and it suffices to explain the travel of genes over

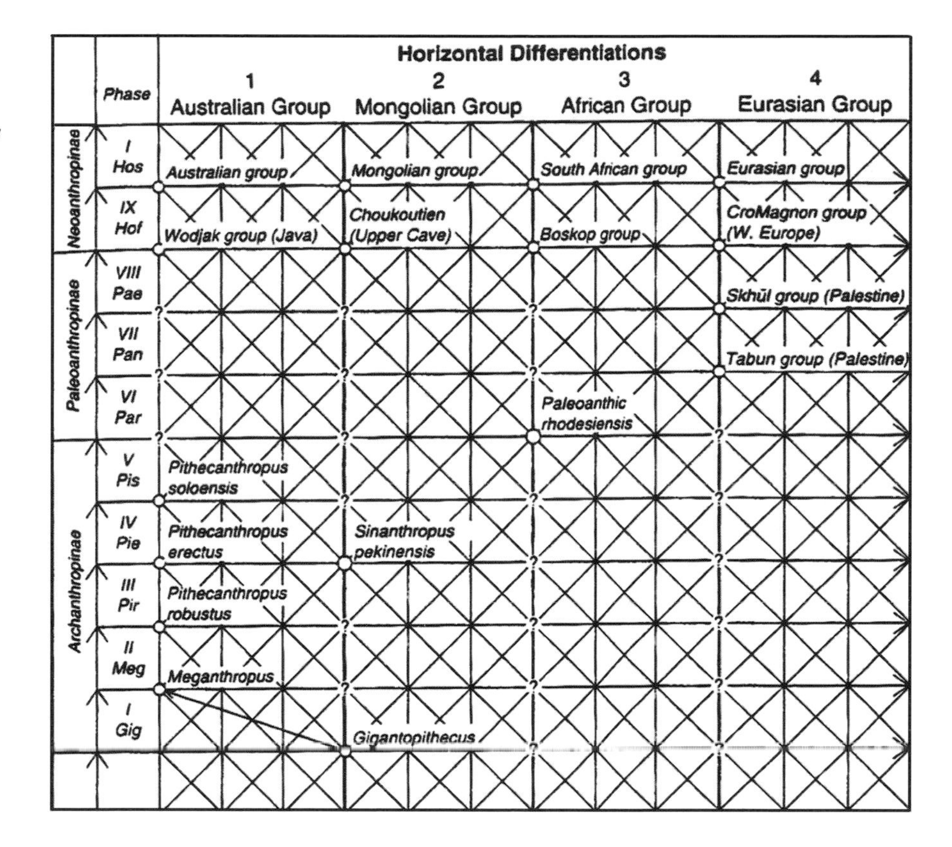


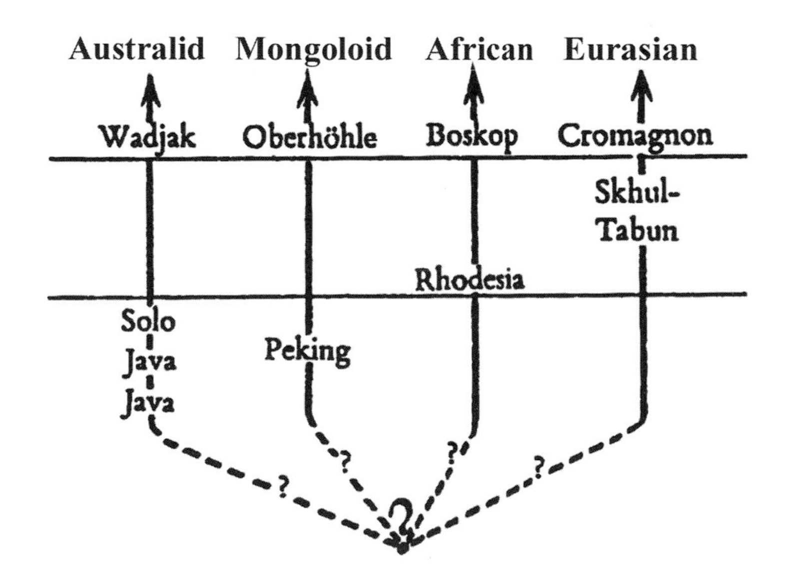


*Figure 57. Above Weidenreich's original trellis model of hominin evolution, which has been much misunderstood in Anglophone Pleistocene archaeology because the diagonal lines were not noticed. Below Howells' false interpretation of Weidenreich's model.*

10,000 km in as few as 100 generations. Yet the enormous time scale available for the development of 'Moderns' amounts to over 2000 generations. Thus the mosaic of human populations of the Final Pleistocene is the result of introgressive hybridisation across contiguous populations subjected to minor demographic adjustments. The catastrophism of the Eve model, by contrast, demands firstly that a tiny population evolved in *complete* genetic isolation for hundreds of millennia in sub-Saharan Africa to the point where it could no longer breed with other humans. This is already demographically absurd. Secondly it perceives some kind of bottleneck in which hominins became almost extinct in Africa. Genetic bottlenecks, however, tend to diminish fitness in the population (Bryant et al. 1986), rather than bring about the population's 'supremacy' (cf. Hawks et al. 2000), as the Eve model demands (consider endemic insular populations). Moreover, in the cases of real catastrophes of the past, such as the Toba and Campagnian Ignimbrite eruptions, no significant changes in technology or culture are indicated (Fedele et al. 200; Fedele and Giaccio 2007).

A second taphonomic flaw in the Eve model is that it assumes the archaeological and paleoanthropological record we have of the Pleistocene is representative. Metamorphology shows this to be a fallacy. If the presumably more sedentary coastal populations in Europe had been more gracile than the more mobile tribes of the hinterland — the only ones we can have any evidence of — our perspective of Ice Age humans is necessarily so distorted it can only provide a parody of history. The notion of invading Africans taking over Europe is as likely to be valid as the account of Noah's Ark: the more one thinks about the logistics, the more absurd it tends to become — except of course for the believers. Any rigorous epistemological review of this model reveals it to be an extremely unlikely demographic hypothesis, based on ideas of 'wandering' tribes and on simplistic notions about the geographical distribution of human fossils. Alternative explanations are much simpler and more parsimonious, and they are even supported by all the available evidence (Bednarik 2008a).

The Replacement Hypothesis has essentially run its course and stands refuted. It will be replaced by another model, but unfortunately this will also be a falsity, whatever alternative it may offer. After 180 years of this vaudeville theatre one would have to be very optimistic to expect that there will be a radical shift in the way Pleistocene archaeology is conducted. What we will see is a rearguard action for a number of years, protestations that Eve advocates have been misunderstood and misinterpreted, and we will end up with yet another faddish hypothesis. And this, too, is a testable proposition.

## REFERENCES

Adcock, G. J., E. S. Dennis, S. Easteal, G. A. Huttley, L. S. Jermiin, W. J. Peacock and A. Thorne 2001. Mitochondrial DNA sequences in ancient

Australians: implications for modern human origins. *Proceedings of the National Academy of Sciences of the United States of America* 98(2): 537–542.

Ampoulange, U. and R. C. Pintaud 1955. Une nouvelle gravure de la grotte de La Grèze (Dordogne). *Bulletin de la Société Préhistorique Française* 52: 249–51.

Anati, E. 1976. *Evolution and style in Camunian rock art.* Archivi 6, Capo di Ponte.

Ayala, F. J. 1996. Response to Templeton. *Science* 272: 1363–1364.

Barinaga, M. 1992. 'African Eve' backers beat a retreat. *Science* 255: 686–687.

Bednarik, R. G. 1989. Perspectives of Koongine Cave and scientistic archaeology. *Australian Archaeology* 29: 9–16.

Bednarik, R. G. 1989b. On the Pleistocene settlement of South America. *Antiquity* 63: 101–11.

Bednarik, R. G. 1990–91. Epistemology in palaeoart studies. *Origini* 15: 57–78.

Bednarik, R. G. 1994. On the scientific study of palaeoart. *Semiotica* 100(2/4): 141–168.

Bednarik, R. G. 1998. The technology of petroglyphs. *Rock Art Research* 15: 23–35.

Bednarik, R. G. 2004. Public archaeology and political dynamics in Portugal. *Public Archaeology* 3(3): 162–166.

Bednarik, R. G. 2008a. The mythical moderns. *Journal of World Prehistory* 21(2): 85–102.

Bednarik, R. G. 2008b. The domestication of humans. *Anthropologie* 46(1): 1–17.

Bednarik, R. G. 2010. Australian rock art of the Pleistocene. *Rock Art Research* 27: 95–120.

Bednarik, R. G. 2011. *The human condition.* Springer, New York.

Bednarik, R. G. 2013. The origins of modern human behavior. In R. G. Bednarik (ed.), *The psychology of human behavior*, pp. 1-58. Nova Science Publishers, New York.

Bednarik, R. G. and M. Kuckenburg 1999. *Nale Tasih: eine Floßfahrt in die Steinzeit.* Jan Thorbecke, Stuttgart.

Bednarik, R. G., G. Kumar, A. Watchman and R. G. Roberts 2005. Preliminary results of the EIP Project. *Rock Art Research* 22: 147–197.

Bräuer, G. 1984. The 'Afro-European sapiens hypothesis' and hominid evolution in East Africa during the late Middle and Upper Pleistocene. In P. Andrews and J. L. Franzen (eds), *The early evolution of man, with special emphasis on Southeast Asia and Africa*, pp. 145–165. Volume 69, Courier Forschungsinstitut Senckenberg.

Brookfield, J. F. Y. 1997. Importance of ancestral DNA ages. *Nature* 388: 134.

Bryant, E. H., S. A. McComas and L. M. Combs 1986. The effect of an experimental bottleneck on quantitative genetic variation in the housefly. *Genetics* 114: 1191–1211.

Caldwell, D. 2008. Are Neanderthal portraits wrong? Neanderthal adaptations to cold and their impact on Palaeolithic populations. *Rock Art Research* 25: 101–116.

Cannon, W. J. and M. J. Ricks 1986. The Lake County, Oregon, rock art inventory: implications for prehistoric settlement and land use patterns. *Contributions to the Archaeology of Oregon* 3: 1–23.

Capitan, L., H. Breuil and D. Peyrony 1912. Les gravures sur cascade stalagmitique de la grotte de la Mairie à Teyat (Dordogne). *Congrès International d'Anthropologie et d'Archéologie Préhistorique, 16e session*, pp. 498–514. Geneva.

Chippindale, C. 2001. Studying ancient pictures as pictures. In D. S. Whitley (ed.), *Handbook of rock art research*, pp. 247–272. AltaMira Press, Walnut Creek, CA.

Chippindale, C. and P. S. C. Taçon (eds) 1998. *The archaeology of rock-art.* Cambridge University Press, Cambridge.

Consens, M. 2006. Between artefacts and egofacts: the power of assigning names. *Rock Art Research* 23: 79–83.

Crivelli Montero, E. A. and M. M. Fernández 1996. Palaeoindian bedrock petroglyphs at Epullán Grande Cave, northern Patagonia, Argentina. *Rock Art Research* 13: 112–17.

Daleau, F. 1896. Les gravures sur roche de la caverne de Pair-non-Pair. *Actes de la Société Archeologique de Bordeaux* 21: 235.

David, P. 1934. Abri de la Chaire à Calvin. *Congrès Préhistoire de France, 11e session,* pp. 372–378. Périgueux.

Draganski, B., C. Gaser, V. Bush, G. Schuierer, U. Bogdahn and A. May 2004. Changes in grey matter induced by training. *Nature* 427(6972): 311–312.

Fedele, F. G. and B. Giaccio 2007. Paleolithic cultural change in western Eurasia across the 40,000 BP timeline: continuities and environmental forcing. In P. Chenna Reddy (ed.), *Exploring the mind of ancient man. Festschrift to Robert G. Bednarik*, pp. 292–316. Research India Press, New Delhi.

Fedele, F. G., B. Giaccio, R. Isaia and G. Orsi 2002. Ecosystem impact of the Campanian Ignimbrite eruption in Late Pleistocene Europe. *Quaternary Research* 57: 420–24.

Garrigan, D., Z. Mobasher, T. Severson, J. A. Wilder and M. F. Hammer 2005. Evidence for archaic Asian ancestry on the human X chromosome. *Molecular Biological Evolution* 22: 189–192.

Goren-Inbar, N. 1986. 1986. A figurine from the Acheulian site of Berekhat Ram. *Mi'Tekufat Ha'Even* 19: 7–12.

Green, R. E., J. Krause, A. W. Briggs, T. Maricic, U. Stenzel, M. Kircher, N. Patterson, H. Li, W. Zhai, M. Hsi-Yang Fritz, N. F. Hansen, E. Y. Durand, A.-S. Malaspinas, J. D. Jensen, T. Marques-Bonet, C. Alkan, K. Prüfer, M. Meyer, H. A. Burbano, J. M. Good, R. Schultz, A. Aximu-Petri, A. Butthof, B. Höber, B. Höffner, M. Siegemund, A. Weihmann, C. Nusbaum, E. S. Lander, C. Russ, N. Novod, J. Affourtit, M. Egholm, C. Verna, P. Rudan, D. Brajkovic, Ž. Kucan, I. Gušic, V. B. Doronichev, L. V. Golovanova, C. Lalueza-Fox, M. de la Rasilla, J. Fortea, A. Rosas, R. W. Schmitz, P. L. F. Johnson, E. E. Eichler, D. Falush, E. Birney, J. C. Mullikin, M. Slatkin,

R. Nielsen, J. Kelso, M. Lachmann, D. Reich and S. Pääbo 2010. A draft sequence of the Neandertal genome. Science 328: 710–722.

Gutierrez, G., D. Sanchez and A. Marin 2002. A reanalysis of the ancient mitochondrial DNA sequences recovered from Neandertal bones. *Molecular Biological Evolution* 19(8): 1359–1366.

Hale, H. M. and N. B. Tindale 1930. Notes on some human remains in the lower Murray valley, South Australia. *Records of the South Australian Museum* 4: 145–218.

Harpending, H. C., M. A. Batzer, M. Gurven, L. B. Jorde, A. R. Rogers and S. T. Sherry 1998. Genetic traces of ancient demography. *Proceedings of the National Academy of Sciences of the United States of America* 95: 1961–1967.

Hardy, J., A. Pittman, A. Myers, K. Gwinn-Hardy, H. C. Fung, R. de Silva, M. Hutton and J. Duckworth 2005. Evidence suggesting that *Homo neanderthalensis* contributed the H2 *MAPT* haplotype to *Homo sapiens*. *Biochemical Society Transactions* 33: 582–585.

Hawks, J., S.-H. Lee, K. Hunley and M. Wolpoff 2000. Population bottlenecks and Pleistocene human evolution. *Molecular Biological Evolution* 17: 2–22.

Helvenston, P. A. 2013. Differences between oral and literate cultures: what we can know about Upper Paleolithic minds. In R. G. Bednarik (ed.), *The psychology of human behavior*, pp. 59-110. Nova Science Publishers, New York.

Kidd, K. K., J. R. Kidd, S. A. Pakstis, C. M. Tishkoff, C. M. Castiglione and G. Strugo 1996. Use of linkage disequilibrium to infer population histories. *American Journal of Physical Anthropology*, Supplement 22: 138.

Lalanne, G. and H. Breuil 1911. L'abri sculpté du Cap-Blanc à Laussel, Dordogne. *L'Anthropologie* 21: 385–402.

Lemozi, A. 1920. Peintures et gravures paléolithiques découvertes dans les grottes des communes d'Espagnac, de Sainte-Eulalie et de Cabrerets. *Bulletin de la Société Préhistorique Française* 17: 256–63.

Maguire, E. A., D. G. Gadian, I. S. Johnsrude, C. D. Good, J. Ashburner, R. S. J. Frackowiak and C. D. Frith 2000. Navigation-related structural change in the hippocampi of taxi drivers. *Proceedings of the National Academy of Sciences, USA* 97(8): 4398–4403.

Mulvaney, J. D. 1969. *The prehistory of Australia*. Thames and Hudson, London.

Norrman, R. 1997. Wolf Cave - Varggrottan - Susiluola; a pre-Ice Age archaeological find in Lappfjärd, Finland. *Studia Archaeologica Ostrobothniensia* 1993–1997. Vasa (in Swedish).

Passemard, E. 1944. La caverne d'Isturitz en pays basque. *Préhistoire* 13: 1–95.

Pavlov, P., J. I. Svendsen and S. Indrelid 2001. Human presence in the European Arctic nearly 40,000 years ago. *Nature* 413: 64–67.

Pennisi, E. 1999. Genetic study shakes up Out of Africa Theory. *Science* 283: 1828.

Pessis, A.-M. 1999. The chronology and evolution of the prehistoric rock paintings in the Serra da Capivara National Park, Piauí, Brazil. In M.

Strecker and P. Bahn (eds), *Dating and the earliest known rock art*, pp. 41–47. Oxbow Books, Oxford.

Pruvost, M., R. Schwarz, V. Bessa Correia, S. Champlot, S. Braguier, N. Morel, Y. Fernandez-Jalvo, T. Grange and E.-M. Geigl 2007. Freshly excavated fossil bones are best for amplification of ancient DNA. *Proceedings of the National Academy of Sciences USA* 104(3): 739–744.

Roberts, R. G., M. Bird, J. Olley, R. Galbraith, E. Lawson, G. Laslett, H. Yoshida, R. Jones, R. Fullagar, G. Jacobsen and Q. Hua 1998. Optical and radiocarbon dating at Jinmium rock shelter in northern Australia. *Nature* 393: 358–62.

Rosenfeld, A., D. Horton and J. Winter 1981. *Early Man in north Queensland.* Terra Australis No. 6, Australian National University, Canberra.

Russell, B. 1959. *The problems of philosophy.* New York.

Saint Mathurin, S. de 1975. Reliefs magdaléniens d'Angles-sur-l'Anglin (Vienne). *Antiquités Nationales* 7: 24–31.

Schulz, H.-P. 2002. The lithic industry from layers IV–V, Susiluola Cave, Western Finland, dated to the Eemian interglacial. *Préhistoire Européenne* 16–17: 7–23.

Schulz, M. 2004. Die Regeln mache ich. *Der Spiegel* 34(18 August): 128–131.

Searle, J. R. 1995. *The construction of social reality.* Allen Lane, London.

Smail, L. M. 2007. *On deep history and the brain.* University of California Press, Berkeley.

Steinbring, J., E. Danziger and R. Callaghan 1987. Middle Archaic petroglyphs in northern North America. *Rock Art Research* 4: 3–16.

Strauss, E. 1999. Can mitochondrial clocks keep time? *Science* 283: 1435–1438.

Swartz, B. K. 1997. An evaluation of rock art conservation practices at Foz Côa, northern Portugal. *Rock Art Research* 14: 73–75.

Templeton, A. R. 1996. Gene lineages and human evolution. *Science* 272: 1363.

Templeton, A. 2002. Out of Africa again and again. *Nature* 416: 45–51.

Templeton, A. R. 2005. Haplotype trees and modern human origins. *Yearbook of Physical Anthropology* 48: 33–59.

Thackeray, A. I., J. F. Thackeray, P. B. Beaumont and J. C. Vogel 1981. Dated rock engravings from Wonderwork Cave, South Africa. *Science* 214: 64–67.

Vigilant, L., M. Stoneking, H. Harpending, K. Hawkes and A. C. Wilson 1991. African populations and the evolution of human mitochondrial DNA. *Science* 253: 1503–1507.

Webb, S. 2006. *The first boat people.* Cambridge Studies in Biological and Evolutionary Anthropology Series 47, Cambridge University Press, Cambridge.

Weiss, M. L. and A. E. Mann 1978. *Human biology and behavior: an anthropological perspective.* Little, Brown and Co., Boston.

# 8

# CONTINGENCIES IN PLEISTOCENE ARCHAEOLOGY

## Frameworks

*Epistemology* (from the Greek *episteme*, 'knowledge', and *logos*, 'theory'), or *the theory of knowledge*, is the branch of philosophy that deals with the nature and origins of knowledge. It addresses standards or norms for justification and reasoning (including logic and probability theory), ideals of rationality, and the effects of specific philosophies (e.g. empiricism, relativism), among other things. Specific canons of rationality are thought to be time-dependent (Lewis 1929: 253; Mannheim 1929-36: 57; Collingwood 1940: Ch. 6; Laudan 1977: 187) as well as culture-specific (Winch 1970: 97), and some authors have even defined them as androcentric conflations biasing science in favour of male ways of experiencing the world. Descriptive epistemic relativism (e.g. deductive inference, causal reasoning; Swoyer 2002) has been improved in recent decades, but remains controversial. As historically and culturally situated creatures we cannot easily step outside our concepts, standards and beliefs to appraise their fit with some mind-independent reality of Kantian 'things-in-themselves'. The trap of extreme relativism, already convincingly opposed by Plato (in his *Theatetus*) can also be avoided by normative epistemic relativism. It holds that while there are no framework-independent facts about the veracity of inference, justification or rationality, there are facts about these variables relative to particular frameworks. Extreme relativism, on the other hand, invites solipsism: if one and the same thing can be *true relative to one framework* and *false relative to another, true for* some groups and *false for* others, there is no truth measure. This was countered by Plato (Figure 58) thus: either the claim that truth is relative is true absolutely or else it is

*Figure 58. Plato (428/427–348/347 BCE).*

only true relative to some framework. If it is true absolutely, then at least one truth is not merely true relative to a framework, rendering the proposition apparently refuted.

A number of philosophers and social scientists (e.g. Quine 1960; Hollis 1967; Davidson 1984) have argued that we can only understand or interpret others if they largely agree with us about what is true, reasonable, justified, or the like. The academic endeavour has resulted in a variety of schools, the disciples of which are separated by "logical gaps": "They think differently, speak a different language, live in a different world" (Polanyi 1958: 151). Or to quote Kuhn:

> In a sense that I am unable to explicate further, the proponents of competing paradigms practice their trades in different worlds. ... Practicing in different worlds, the two groups of scientists see different things when they look from the same direction (Kuhn 1970: 150).

Some of these branches of the academic project have chosen to operate under a collective umbrella framework, called *science*; others have developed their own various frameworks. Science, today, favours a normative epistemic relativism over an over-simplified absolutism, but demands specific procedures of refutation and repeatability of experiments, and strives for refutable theories cast in terms of causes. After all, quantum theory implies that determinism fails and that objects need not always have determinate locations in space and time or determinate magnitudes (like a particular momentum or energy or spin). In all of this, the issue of testability of hypotheses is utterly paramount, involving two components: first, the logical property that is variously described as contingency, defeasibility or falsifiability (which means that counterexamples to the hypothesis are logically possible); and second, the practical feasibility of observing a reproducible series of such counterexamples if they do exist. Thus a hypothesis is testable if there is some real hope of deciding whether it is true or false of real experience. Yet the principal epistemological characteristic of archaeology as it has been conducted until now is its poor refutability. Relativism decrees that this does not render archaeology in some way inferior; archaeology is simply an epistemic framework that has chosen to eschew scientific demands in favour of a different framework.

It is perhaps in response to this non-scientific base that the discipline has developed a preference for authority. It is widely considered inappropriate to challenge its upper echelons, simply because to undermine their eminence would impact on the credibility of the discipline. As they are the columns holding up its structure it is essential for its survival that they not be subjected to doubts. The principle is well expressed in countless reactions to challenges, of which I cited just two in Chapter 5. Archaeologist Dorothy Garrod, when caught attempting to salt the site of Glozel in France to discredit the discovery, first denied this, but when confronted by several witnesses admitted the act, and years later confessed that she had done this "for the honour of the discipline". The careers and reputations of many people involved were at stake (beginning with that of her mentor, Henri Breuil), and they were more important than the archaeology of the Glozel site or the fate of its discoverer, Émil Fradin. His life was at the time being destroyed by those whose academic reputations were at stake because of what he had unwittingly uncovered. This is of relevance to understanding the epistemology of the discipline.

The same attitude, that the credibility of professional archaeologists is more important than the veracity of their propositions, can be identified in countless other episodes in the history of the discipline. A more recent example, from my experience, was the reaction of the principal protagonist opposing scientific dating of the rock art in the Côa valley of Portugal. When I expressed my support for blind tests, he retorted in 1995 in the journal *Antiquity* that these were disrespectful, that one should have "consideration for colleagues" and that blind tests were "unethical". This illustrates not only the incommensurability gap (Whorf 1956; Feyerabend 1962; Kuhn 1970) between archaeology and science; it also manifests the self-corrupting paradigm that finds it is more considerate to allow the deception of colleagues and the public. Presumably it is also 'disrespectful' and 'unethical' to falsify the propositions of colleagues. The Portuguese professor even stated in 2001 that criticism of him "serves to create confusion, and boosts a rejection of archaeology by the media and the public — 'those folks who never know what exactly it is they want and are fighting each other anyway' ". This dimension of archaeology, its social construction (Berger and Luckmann 1966) of placing credibility above veracity, requires detailed analysis and exposure, because it defines its epistemology and explains the treatment of both dissenters and amateurs. They not only *may* be disproved, they *must* be disproved — even if that were to involve salting a site with fake objects.

On the issue of archaeology's status as a science, as a refutable system of knowledge claims, archaeologists are themselves very much divided. Those who favour post-processualism or post-modernism consider that the discipline is no science, and often even state that it does not need to be one. Many others, in an emotional need to see their discipline as a science, promote a cargo-cult-like scientism: if scientific data are imported it might become a science. Some have argued that archaeology is not fundamentally different from, say, geology. For instance in countering the contention that excavated and destroyed strata cannot be subjected to testing, one argument fielded is that when geology (or any of a number of scientific disciplines) removes core samples, it is similar to the excavation of archaeological sediments. This argument illuminates the relevant epistemologies. When a core of ice, sediment or rock is taken, a relatively homogenous deposit is sampled to gain an understanding of its composition, stratigraphy or other properties. If a second core hole were drilled next to the first, the results would be expected to be identical; hence the process is repeatable. This differs significantly from the project of archaeological excavation, where every square of sediment is expected to be different, have different contents (artefacts, interments, structures, occupation floors etc.) and properties. If that were not the case there would be no point excavating. So each excavation unit is unique.

Therefore the difference between archaeology and, say, geology could not be more fundamental. In geology, the extent of the resource (the ore body, oil field, or whatever) is mapped without exposing it. In archaeology the resource is exposed and destroyed completely, through the excavation of the part of a site explored. Once the sediment has been removed we only have the records: section and plan drawings, finds saved, and sediment samples taken. Most of what the archaeologist says about the resource, the 'cultural' deposit, is no longer falsifiable. We can only accept the report on authority, which in science is unacceptable. Records such as section drawings are merely artistic impressions; they are not hard evidence. As considered in Chapter 7, if we commissioned ten archaeologists to independently draw sections of a complex stratigraphy, we would elicit ten different drawings. This could be easily tested, but as noted, archaeologists are averse to such 'blind tests', which some of them consider to be "disrespectful to colleagues". Yet most archaeologists do not conduct even simple scientific tests on site during excavations, and therefore their pronouncements of what they saw in the excavation can be questioned. The fact that their determinations cannot be tested, cannot be falsified, is therefore a fundamental concern.

## Social realities

Our next epistemic encumbrance is the fact that the crucial common denominator (CCD; first proposed in Bednarik 1990–91) of phenomenon categories is difficult enough to determine in today's world, but is probably impossible to identify for past cultural systems. Today's objects in our perceptible world do not exist independent of conceptual frameworks (Putnam 1981: 52).

> If, as I maintain, 'objects' themselves are *as much made as discovered*, as much products of our conceptual *invention* as of the 'objective' factor in experience, the factor independent of our will, then of course objects intrinsically belong under certain labels because those labels are just the *tools we use to construct a version* of the world with such objects in the first place (Putnam 1981: 54, his emphases).

Examples would include biological species (which are often debatable categories, even though we agree that viable reproductive ability is their CCD) or rock types (which often lack an objective taxonomy as they are manifestations of compositional continuums). To then extend an epistemologically questionable practice to an epistemologically challenged field such as Pleistocene archaeology is to court disaster. Consider, for instance the intricate stone tool nomenclatures we have invented, on which our hypothetical cultural categories depend, and on whose veracity much of Pleistocene archaeology stands and falls. Does anyone seriously believe that these entirely arbitrary and etic stone tool types are real? Do we believe that one hypothetical 'Aurignacian' (there is no proof that such a group actually existed, as a tribe, nation or ethnic entity) person said to another: "Pass me the keeled scraper, this waisted blade is unsuitable for making this wooden thingy"? Or that he exclaimed: "Look what a lovely Acheulian hand-axe I just found in the streambed! I can knap it into at least six of those Abri Audi points that have been so fashionable lately." All of the designations and taxonomic units of Pleistocene archaeology are simply inventions (cf. Clark 2009: 26), of relevance in one framework, false in many others, and almost certainly false in the cognitive framework of the people of the time when the tools were actually made or used:

> Quite literally, men of those days lived in a different world because their instruments of intellectual interpretation were so different (Lewis 1929: 253).

In this sense archaeology differs from all other disciplines, because they all endeavour to define real entities, whereas in Pleistocene archaeology there should not even be pretence of that. Searle's (1995) illumination of social realities distinguishes between the brute facts of an object's intrinsic characteristics and those that are observer-relative, or 'institutional' facts. For instance, an object may be made partly of wood, partly of metal. Its property of being a screwdriver exists only because the person who makes or uses it represents it as such. Precisely the same applies to an object made in the Pleistocene; it has factual properties, and socially constructed, observer-relative properties. However, there is no evidence that the latter are shared between the ancient maker and user of the object, and its modern-day archaeologist interpreter. The term *déformation professionnelle* refers to this issue: professional training, which as a process mimics the neuropsychology of obsessive compulsive disorder (Bednarik 2011), also results in a distortion in the way the world is perceived. Confirmation bias (Wason 1960; Evans et al. 1983) can only add to the sophistry. Observer-relative definitions, attributions and claims about the distant human past are clearly not in themselves of scientific utility; they need to be subjected to metamorphological analysis, which so far has not occurred in a systematic fashion.

But again, this does not automatically show that all of Pleistocene archaeology is nonsense; it merely indicates that, on the basis of reasonable probability, a certain proportion should be assumed to be false. According to Kuhn (2000: 30), scientific revolutions occur through "change in several

*Figure 59. Palaeolithic zoomorphs, Lascaux.*

of the taxonomic categories prerequisite to scientific descriptions and generalisation". However, correcting taxonomic categories in Pleistocene archaeology would not be an easy task. Firstly, there is the intransigence of the discipline to contend with; and secondly, how would one set about creating a superior taxonomy? This, of course, would be the subject of a separate book. Here we are only concerned with the much less ambitious project of explaining how metamorphology would need to approach these matters.

To select from the myriad misconceptions of Pleistocene archaeology just one for a representative analysis, let us consider the common conception that Palaeolithic cave art is dominated by zoomorphs (Figure 59). Bearing in mind that animal images from that corpus are in fact outnumbered four or five times by apparently non-figurative motifs, the question arises why there should be a conception among the public, and even among archaeologists, that animal images prevail numerically. There are many palaeoart traditions around the world whose iconography is dominated by zoomorphs, so this is not at all a variable (CCD) defining the Franco-Cantabrian cave art. But by far the most specific type of motifs that are exclusive to the Upper Palaeolithic corpus are the so-called signs, which are repeated many times (Figure 60).

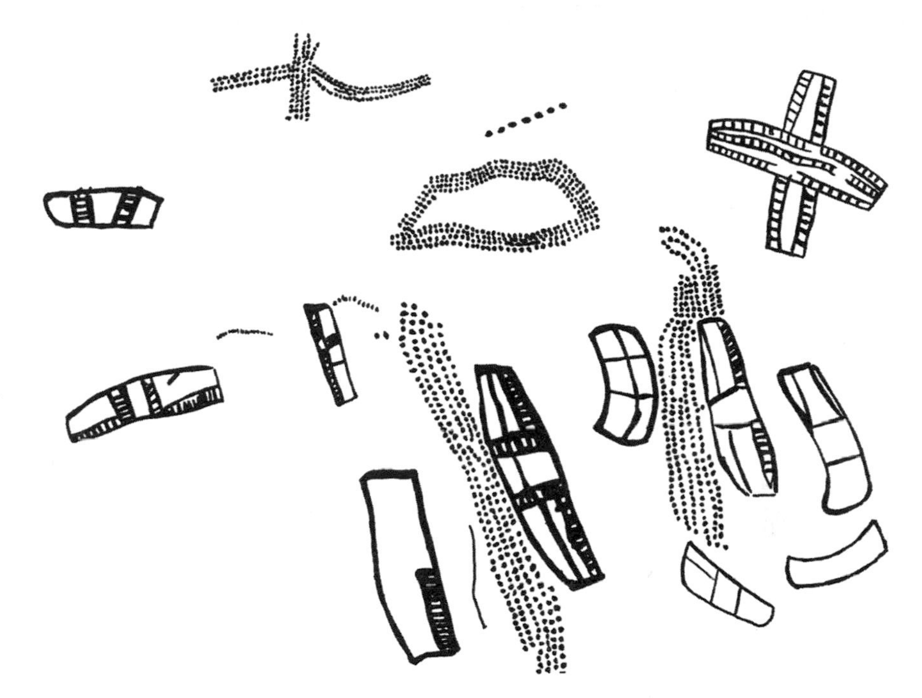

*Figure 60. Example of non-figurative Palaeolithic rock art.*

Their meaning remains unknown and they have attracted far less attention than the zoomorphs. These 'signs' occur in no other rock art tradition; hence they are reliable cultural markers. And yet there is a perception among archaeologists that zoomorphs are the prime indicator of Palaeolithic rock art. This has led to many unsuccessful searches for Pleistocene rock art across Eurasia and even in North America, including a tendency to label any animal imagery in Eurasian rock art as having to be of the Palaeolithic (Breuil 1952; Okladnikov 1959; Mori 1974; Balbín et al. 1991; Kohl and Burgstaller 1992; Molodin and Cheremisin 1993; Zilhão et al. 1997; Bahn et al. 2003).

How did this misconception arise? While it is obvious that publications tend to overemphasise the figurative content of this cave art, more subtle factors may also be at work here. Judging from the response patterns of visitors of publicly accessible cave sites, the public prefers that part of the art it believes it 'can identify', and has rather less interest in 'un-interpretable signs'. This resembles the reaction of children when viewing rock art panels. It seems that either researchers react in the same fashion and form priorities on that basis, or they are good judges of what the public appreciates and simply adapt their priorities accordingly. While this preoccupation no doubt curries favour with the public, it is detrimental to research. From a scientific perspective, the non-figurative component is likely to be the more important. As the apparent part of a communication system it is much more culture-specific than the invented 'styles' of researchers, because figurative images appear to communicate across cultures — which is precisely the point. So we tend to ignore the culture-specific motifs in favour of what we think we can comprehend (in spite of being historically and culturally situated creatures that cannot correctly identify alien iconographies), and have instead invented 'styles' of animal images as a means of defining 'cultural identities'. Meanwhile, the complex communication systems of the Upper Palaeolithic remain unexplored, even though we know that all symbol systems (be they computer languages, conventions for diagrams, styles of painting) influence perception and thought (Goodman 1978; cf. Malafouris 2008). In short, archaeology fails to clarify; it only serves to muddle issues that would be much better left to semiotic study.

This example using the incidence of iconography is only one of many I could choose from. In Chapter 6 we began exploring the generic reasons why paradigms of Pleistocene archaeology essentially have to amount to distortions of what really happened in history. There are myriad factors, to be untangled by metamorphology, mindful of the contingency of our modes of thought and evaluation. The most parsimonious system of modern science, normative epistemic relativism, concedes the lack of framework-independent facts about general veracity, but preserves the veracity of inference, justification or rationality relative to specific frameworks. Therefore the knack is to rank the frameworks themselves according to their validity or credibility. The relativist admits that he inhabits one of them, and thus acknowledges that his claims are only true relative to it. The extreme relativist, or postmodernist, courts a solipsism capable of accepting anything because "everything is only a social construction".

The normative epistemic relativist, then, needs to ask: what is the relative epistemic strength of the received knowledge of Pleistocene archaeology as a framework of knowledge claims? This, of course, is the subject I seek to address in this volume, and it has become evident that we have inherited a framework that is so poorly equipped it might for some observers resemble a religion. In comparing it to some of the best-performing disciplines, this field needs to be judged by both its historical performance and its muddled epistemology.

## Towards an epistemology of Pleistocene archaeology

As an epistemologist of archaeology I investigate where the knowledge archaeologists believe they possess originates, and what its intrinsic nature is. If I were to do the same in a field of science, such analyses would be welcomed as providing useful testing of frameworks. But in archaeology, because of its great dependence upon authority, such attention is not only unwelcome, it is vigorously discouraged, and it can incur the wrath of much of the discipline. This is in fact one of the most fascinating epistemological facets of the field. While detrimental or restrictive procedures are encouraged, through a whole raft of practices (e.g. to be admitted to a lucrative club of archaeological consultants, one is required to be nominated from within, i.e. by existing members), critical analysis is frowned upon. Archaeology simply does not wish to confront its epistemology, and it implicitly rejects the principles of a meritocracy.

Should such a discipline be supported by society, by the public and by public institutions? If we consider the primarily political content and thrust of the discipline, its neocolonialist basis and nature, its practices of destroying cultural heritage sites, and its unhealthy curatorial aspirations and monopoly-forming practices — what, precisely, are the benefits archaeology offers society? The endless series of blunders over the last couple of centuries, the always controversial 'explanations' of the past are not persuasive factors. The books and television films for the public may be good entertainment, but their frequent lack of veracity warrants improvements, not endorsement. On balance, one might be tempted to advocate the discipline's closure. Or, perhaps, that it should revert to its previous state, a hobby or scholarly interest of gentlefolk researchers, people who lack the missionary zeal of modern Pleistocene archaeology — and who have, as history shows, performed rather better than their modern 'professional' counterparts.

But a thorough epistemological analysis of the field also shows that many researchers in Pleistocene archaeology do make extraordinary efforts to provide quality work. Moreover, in spite of the many shortcomings, the discipline would in my view have the potential to improve, and to improve quite considerably. I have canvassed some of the potential improvements, especially a destruction of dogmas and their replacement with weak propositions of falsifiable formats. Many of the claims Pleistocene archaeology has so far presented were precipitate, and the picture we have of Ice Age hominins is so profoundly inadequate because the kinds of models that grow, mushroom-like, in this dark are often only designed to create reputations and careers (Henneberg and Schofield 2008). It is this system that needs to be dismantled, not the discipline per se. It needs to be fully appreciated that, in a few centuries hence, our present knowledge of our Pleistocene ancestors will look just as inconsequential as such knowledge 200 years ago appears today. It is from that historical perspective that we need to approach this subject. And we need to appreciate why it is that Pleistocene archaeology makes so little progress, or merely stands still, while the hard sciences progress at a breathtaking, relentless pace.

Consider just two of the more recent disciplines, both only half a century old: plate tectonics and ethology. In a matter of decades they progressed from embryonic stages to incredible complexity. Or consider genetics: after its introduction by Mendel (1866) it remained stagnant, if not ignored, for much of a century, but look at its sophistication today! By comparison, Pleistocene archaeology and paleoanthropology are both static: if skeletal remains of a small primate are found on the island of Flores today, the discipline erupts into a cacophony of competing interpretations. This is reminiscent of an astronomy that, in our time, still argues about whether the Sun or the Earth rotates around the other. As we saw in Chapter 5, anyone can judge that the Flores creature was a primate, and if the experts of the world form opinions representing the entire possible range, it becomes starkly apparent what happens when authority in a non-falsifiable discipline is challenged. Not only do we now have a proposal that the specimen may have been planted in the sediment, presumably as part of an elaborate hoax (Henneberg and Schofield 2008), as well as the documented academic misconduct and 'skull-duggery' relating to this discovery; there is also the spectacle of a bitter battle between competing schools of thought. This is precisely the same pattern that marred previous palaeoanthropological discoveries.

The question arising from such observations is: what is it that renders Pleistocene archaeology so accident prone and unreliable? Today I am probably the most avid debunker of archaeological blunders and mistakes, and I have been asked whether there is a system in how I choose foci of interest. The answer is that taphonomic logic tends to identify the most mistake-prone areas quite readily, and it is then merely a question of homing in on false claims on the basis of data. But there is one other factor, which is related to a false logic implicit in the extreme conservatism of the discipline. Because of the impossibility of subjecting most archaeological claims to systematic refutation, this extreme conservatism has developed as a defence system. It led to a reliance on, and preference for, authority, which is itself already an epistemological impairment. But more relevantly, it fostered a specific brand of minimalist dogmas. These are based on the assumption that to protect the paradigm against unsound challenges to what provides a 'semblance of certainty', it is best to resist changes to a dogma. The more we resist, the closer the process resembles an inverted form of falsificationism. In other words, if the dogma says humans at a certain time had acquired a specific technology or ability, any data supporting an earlier introduction has to be resisted strenuously, until the evidence becomes overwhelming. This protects the received knowledge against frivolous claims, which have to be opposed vigorously. Claims that do not challenge the dogma, on the other hand, could be readily accepted, irrespective of whether they are false; they could not damage the dogma.

This provides a key to understanding the false epistemology of Pleistocene archaeology. It shows that anything can be proposed and will be accepted, provided it leaves the dogma intact: *compliance with dogma, not veracity, is the criterion of acceptance*. But therein lies the problem: the dogma is itself likely to be false. To see this, we need to recall (see previous chapter) how the discipline is entirely dependent upon a random historical sequence of discoveries: if that sequence had been different, our received Pleistocene archaeology also would be very different. To defend a randomly acquired model on no basis other than its historical precedence is demonstrably fallacious. Moreover, the practice of retreating as reluctantly as possible from such an incomplete model is logically unsupportable. It argues for a top-down retreat strategy — a regression from a *contingent* state of limited validity, instead of starting with the null hypothesis that sampling errors are inherent in all archaeological work.

A better alternative than adherence to a flawed model would be to start with the null hypothesis that our ancestors of around 5 to 8 million years ago were not fundamentally different from chimps or bonobos, and that since then they developed into what we are today. Since we are not certain how this development occurred, it would be judicious to assume (as Haeckel did in the 19th century) that, half way through, we might expect to find creatures about half way between apes and us. In an anatomical sense that is indeed what we are finding, and it is generally agreed that physical evolution of hominins occurred fairly gradually, over the entire period. Similarly, encephalisation (enlargement of the brain) was undeniably a gradual process. Why, then, is it that Pleistocene archaeology assumes that cognitive or intellectual development is almost entirely a feature of the last third of the Late Pleistocene (the last 40,000 years)? Bearing in mind the enormous cost of encephalisation to mothers and whole societies (consequences of the need to expel large skulls through the birth canal, long-term dependency of infants), it is illogical to maintain that these large brains were not used. It is also evolutionary humbug: like any other evolutionary change, organ size must be selected for; it does not just increase randomly. So instead of insisting, as most Pleistocene archaeologists do, that pre-modern hominins were fundamentally primitive, because the dogma demands so, it would be far more realistic to postulate that frontal and parietal lobes were used for thinking in proportion to their size. This makes biological sense, but it is entirely irreconcilable with the model archaeology offers us: that so-called Neanderthals with their 13% larger brains were vastly more primitive than our gracile forebears. Therefore the conundrum seems to be due to the archaeological dogma more than any other factor: it simply does not fit.

But why should one expect the dogma to fit? It is probably false inherently; it is a minimalist interpretation of largely distorted data; it was arrived at by non-random sampling and by inadequate methodologies. Moreover, bearing in mind the effects of taphonomy, the available, intrinsically incomplete record is precisely what is to be expected, especially from the earliest periods of human history.

In other words, the dysfunctional relationship between the models of Pleistocene archaeology as disseminated by the great journals and institutions on the one hand, and those derived by more scientific and more critical approaches (Bednarik 2011a, 2011b) on the other is fully predictable. It illustrates the difference between grandiose story telling and the more sobering probability scenarios of science. There can be no science without the facility of falsification or testability, yet we have seen that there are several fundamental impairments preventing the testing of archaeological propositions:

1. The principal method, excavation of sediments, cannot yield falsifiable knowledge claims, because the resource itself is destroyed in the process. Therefore all claims about what excavation has shown are based on authority alone, which in science is not acceptable.
2. Propositions of archaeology about what happened in the past cannot be falsified directly by other purely archaeological claims (although in many cases they may be susceptible to indirect falsification by scientific methods).
3. Archaeology cannot produce predictive postulates (e.g. about societies that no longer exist) capable of being subjected to testing.
4. Archaeology cannot, with any credibility, cast hypotheses in terms of causal relationships.
5. The classes of nomenclatures or taxonomies invented by archaeology, especially Pleistocene archaeology, are non-falsifiable; they are etic and free-standing constructs. This applies, for instance, to artefacts, motivations, beliefs, intentions, social models and practices.

Concerning item 2, the inability of archaeology to accommodate the canons of science, illustrated by the misuse and misinterpretation of scientific data or modes of discourse, complicates matters greatly. For the purpose of illustrating this point with a randomly selected example (many others come to mind, especially in statistics), consider the use of Voronoi diagrams or Dirichlet tesselation in archaeological theory (although here it is mistakenly referred to as Thiessen polygons, an application of the principles to meteorology). For instance Clarke (1978: Fig. 116) treats Iron Age sites as Voronoi sites, which means that each of these occupation sites has a Voronoi cell consisting of all points closer to that site than to any other. This is in principle nearly correct, but the model's application in archaeology can only give rise to falsities. In Clarke's example of Iron Age sites, the following factors show this:

a. We can never consider all Iron Age sites in a given area, but only those that have survived, and only those that have been located so far.
b. Unless we can consider only those that coexisted, comparing the geometric distribution of all known Iron Age sites would only yield falsities.

This example shows how taphonomic logic and metamorphology immediately debunk an archaeological misapplication of a scientific method that has numerous valid applications in other disciplines. Its denouement can be extrapolated to most practices of Pleistocene archaeology, demonstrating that much of it amounts to a mythology (e.g. Bednarik 1992, 2008, 2011a).

At the end of these deliberations I would like to return to the question I posed in Chapter 1: does archaeology understand its role in needing to explore how the cognitive niche of hominins might have been established?

Does it appreciate this need if we are to bring any light to bear on how human constructs of reality (which in the final analysis determine all epistemology) came into being? And can we confidently rely on this discipline's capability of extracting the kind of information needed in such a quest? All factors considered fairly, I think we would be obliged to point to the extensive list of epistemic deficiencies I have sought to canvass.

## REFERENCES


Bahn, P., P. Pettitt and S. Ripoll 2003. Discovery of Palaeolithic cave art in Britain. *Antiquity* 77: 227–231.

Balbín Behrmann, R. de, J. Alcolea Gonzalez, M. Santonja and R. Perez Martin 1991. Siega Verde (Salamanca). *Yacimiento artístico paleolítico al aire libre*, pp. 33–48. *Del paleolítico a la historia*, Museo de Salamanca, Salamanca.

Bednarik, R. G. 1990-91. Epistemology in palaeoart studies. *Origini* 15: 57–78.

Bednarik, R. G. 1992. Palaeoart and archaeological myths. *Cambridge Archaeological Journal* 2(1): 27–43.

Bednarik, R. G. 2008. The mythical Moderns. *Journal of World Prehistory* 21(2): 85–102.

Bednarik, R. G. 2011a. *The human condition*. Springer, New York.

Bednarik, R. G. 2011b. The origins of human modernity. *Humanities* 1(1), 1–53; http://www.mdpi.com/2076-0787/1/1/1/.

Berger, P. L. and T. Luckmann 1966. *The social construction of reality: a treatise in the sociology of knowledge*. Anchor Books, Garden City, NY.

Breuil, H. 1952. *Four hundred centuries of cave art*. Centre d'Études et de Documentation Préhistoriques, Montignac.

Clark, G. A. 2009. Accidents of history: conceptual frameworks in paleoarchaeology. In M. Camps and P. R. Chauhan (eds), *Sourcebook of Paleolithic transitions: methods, theories, and interpretations*, pp. 19–41. Springer, New York.

Clarke, D. 1978. *Analytical archaeology* (second edn). Methuen, London.

Collingwood, R. G. 1940. *An essay on metaphysics*. Clarendon Press, Oxford.

Davidson, D. 1984. *Inquiries into truth and interpretation*. Clarendon Press, Oxford.

Evans, J., J. L. Barston and P. Pollard 1983. On the conflict between logic and belief in syllogistic reasoning. *Memory and Cognition* 11: 295–306.

Feyerabend, P. 1962. Explanation, reduction and empiricism. In H. Feigl and G. Maxwell (eds), *Minnesota studies in the philosophy of science, Vol 3*, pp. 28–97. University of Minnesota Press, Minneapolis, MI.

Goodman, N. 1978. *Ways of worldmaking*. Hackett Publishing Company, Indianapolis, IN.

Hollis, M. 1967. The limits of irrationality. *Archives Européenes de Sociologie* 7: 265–271.

Kohl, H. and E. Burgstaller 1992. *Eiszeit in Oberösterreich: Paläolithikum-Felsbilder.* Österreichisches Felsbildermuseum, Spital am Pyhrn.

Kuhn, T. 1970. *The structure of scientific revolutions*, 2nd edn. University of Chicago Press, Chicago, IL.

Kuhn, T. 2000. *The road since structure.* University of Chicago Press, Chicago.

Laudan, L. 1977. *Progress and its problems.* University of California Press, Berkeley.

Lewis, C. I. 1929. *Mind and the world order.* Charles Scribners Sons, New York.

*Malafouris, L. 2008. Beads for a plastic mind: the 'blind man stick' (BMS) hypothesis and the active nature of material culture. Cambridge Archaeological Journal 18(3): 401–414.*

Mannheim, K. 1929–36. *Ideologie und utopie.* F. Choen, Bonn.

Mendel, J. G. 1866. Versuche über Pflanzen-Hybriden. *Verhandlungen des naturforschenden Vereines in Brünn* 4: 3–47.

Molodin, V. I. and D. V. Cheremisin 1993. *Drevneĭshie petroglify Altaya.* Obozrenie, Novosibirsk.

Mori, F. 1974. The earliest Saharan rock-engravings. *Antiquity* 48: 87–92.

Okladnikov, A. P. 1959. *Shishkinsie pisanitsi.* Nauka, Irkutsk.

Polanyi, M. 1958. *Personal knowledge.* Routledge, London.

Putnam, H. 1981. *Reason, truth and history.* Cambridge University Press, Cambridge.

Quine, W. V. O. 1960. *Word and object.* MIT Press, Cambridge, MA.

Searle, J. R. 1995. *The construction of social reality.* Allen Lane, London.

Swoyer, C. 2002. Judgment and decision making: extrapolations and applications. In R. Gowda and J. Fox (eds), *Judgments, decisions, and public policy*, pp. 9–45. Cambridge University Press. Cambridge.

Wason, P. C. 1960. On the failure to eliminate hypotheses in a conceptual task. *Quarterly Journal of Experimental Psychology* 12: 129–140.

Whorf, B. L. 1956. *Language. Thought and reality.* M.I.T. Press, Cambridge, MA.

Winch, P. 1970. Understanding a primitive society. In B. R. Wilson (ed.), *Rationality.* Harper & Row, New York.

Zilhão, J., T. Aubry, A. F. Carvalho, A. M. Baptista, M. V. Gomes and J. Meireles 1997. The rock art of the Côa valley (Portugal) and its archaeological context: first results of current research. *Journal of European Archaeology* 5: 7–49.

# 9

# TAXONOMIES IN ARCHAEOLOGY

## Prelude to a summary

So far this book might suggest a preference of science over archaeology, but this issue is rather more complex than it appears. The technocracy we call science is, after all, largely dedicated to anthropocentric ends. Moreover, many of its practitioners are in the employ of political or corporate structures, which affects their independence and questions their credentials. While applied science has become a powerful icon during the 20th century, nearly all research funding is directed towards the ultimately selfish aims of humans themselves — especially to medical, strategic and space research. The first two seem to strive for contradictory ideals: prolonging and saving human life on the one hand, and perfecting ways of destroying it on the other. But in a more objective sense they are really very similar, being both about self-preservation: either at the level of species or individuals, or at the level of political, ethnic or religious entities. Medical research is very good for the human species but it is a self-indulgent, self-centred and generously funded pursuit of self-preservation of one species. In the final analysis it is founded on what Albert Einstein laconically called the "ideal of swine". The pursuit of 'human happiness', after all, is of very peripheral importance to the world, even within this tiny speck in the universe we are vaguely familiar with. One does not have to be 'environmentally aware' to accept that our species has not been a great boon to our planet's well-being. Encouraged by the righteousness of its historical ideologies and religions, the human species has come to regard this planet as its property. This relationship to the Earth and the natural system that renders our existence possible is very different to that of the surviving indigenous peoples of the world. To some extent it is attributable also to intellectually corrupt aspects of science. It may sound provocative to state this, but since it is undeniably true we ought to ask why such a statement should sound unsettling.

The amount of genuinely altruistic science (i.e. not designed to serve human self-interest) being done today is minuscule (and may well have been relatively greater in the 19th century), and is largely restricted to areas that are devoid of any economic potential. A science aimed at the betterment of humanity is not only a very impoverished science; it is a corrupted form of science. To some scholars, the use of the word 'science' in most of the contexts in which it is used in popular vernacular is demeaning. True science must be conducted outside of economic considerations and personal ambitions, separate from its practical significance to one species. To be valid, a scientific proposition or hypothesis must (in addition to being refutable) be acceptable to any conceivable intelligent organism in the universe, irrespective of whether such an organism has ever existed or will ever exist. It must be valid in any conceivable system of intelligence, even one of an organism sharing none of the human sensory faculties, but finding its way in the world with the help of faculties we cannot even begin to imagine. The propositions such hypothetical beings would share with us as being valid are likely to include a number that refer to real reality, a concept of a reality that exists independent of our conceptualisations of the world. Emmanuel Kant calls this *Das Ding an sich*, or 'The thing in itself', and other philosophers have used various definitions and metaphors for it. In Plato's simile of the cave, the noble purpose of science would be to speculate about what is behind the prisoners' backs, using procedures of logic. Naturally we cannot be sure about the validity of any deduction we make in this fashion, which is one of several reasons why refutation is the scientific way to proceed: hypotheses are not here to be proved, they are here to be disproved (Popper 1934). Or, as Ludwig Wittgenstein might say, philosophy is not about discovering truths, but about dissolving human confusions: his "Whereof one cannot speak thereof one must be silent" would be an apt motto for scientists. And on the subject of truth, Nietzsche deserves to be quoted: "Truth is an army of flexible metaphors and anthropomorphisms".

Anthropocentric realities are those conceptualisations hominins have developed of the world, the metaphysical models that so dominate the very functioning of our 'mind' (as the functioning of our nervous system is often called) that we are nearly incapable of thinking outside of them. Their long use over hundreds of millennia has contributed to the way our neural structures developed phylogenetically, and they determine the ways in which we *can* think. This presents the greatest challenge to scientific thought: if we as a species think in ways predetermined by past metaphysical or ontological models, none of us can be free of what the neuroscientist calls magical thinking (and attributes to a lack of integration between the left prefrontal cortical areas and memory). How much can we trust our own empirical judgment? The dominant anthropocentric reality of most of today's humans includes strong concepts of spatiality and time: a clear three-dimensional entity experienced as space, and the non-spatial linear continuum in which events seem to occur in an irreversible succession experienced as time. These, and many other concepts of reality would have been developed in tandem with the evolution of the hominin brain, and when we begin to consider the implications of realities that are possibly predicated, at least in part, on such ontological contributors as neural structures determined by cultural agents, we have every reason for a fundamental scepticism. This is no longer commonplace scientific scepticism, it is a most profoundly felt humility — the result of realising that the gap between our model of the world (i.e. the reality we experience, apparently collectively) and the real world is in all probability very much greater than even the sceptics among us had suspected. Modern 'science' sometimes ignores this frightening hiatus, having become absorbed by its own self-confidence and its techno-toys.

This credibility gap between the false world of anthropocentrism on the one hand, amplified by empiricist and confirmationist 'science', and the abstraction of an objective reality, which is thought to exist, is apparent even at various trivial levels. For instance, the continuing human endeavour to dominate nature, to incessantly manipulate ecology at the material level in niche construction (Laland et al. 2000; Odling-Smee et al. 2003), is a more mundane expression of the human propensity to create our own environments. One could illustrate the point by considering the disastrous and apparently accelerating modifications of specific environments by human societies in many parts of the world. But at the much more fundamental level, early humans have been creating their conceptual world for as long as they have conceptualised about the world, and this idea of the world we have inherited from them is simply a conceptual artefact — a niche. Archaeology has provided no credible information about this dimension. And yet, to be logically consistent, we must go one step further and state the obvious corollary: those parts of the human brain, those neural pathways that have been formed during culturally driven encephalisation (the phylogenic increase in brain volume over several million years), could be considered to be incidental artefacts. I have considered them in some detail elsewhere (Bednarik 2011a). One might argue that, while this is essentially true, the 'artificiality' of neural structures has no conscious derivation, whereas archaeological artefacts, for instance, are the product of conscious contemplation. But this would raise the vexed question of human intentionality, which science has shown itself singularly incapable of dealing with effectively.

What follows from all this is that, if we would know and understand the processes of the cognitive evolution of hominins, we would be able to trace the way hominins developed their ideas about the world, about reality. Not only would this show us why and how past anthropocentric realities were created, it would also tell us why alternative ones were not, and what their consequences would have been had they been chosen instead. The most sensible course of action in any quest to explore the origins of human realities seems to be to explore the early development of human consciousness, focusing on the period during which the cognitive niche of hominins might have been established. *And this is where archaeology enters the arena.* Without it we are not very likely to learn a great deal about the stuff anthropocentrism is made of. The question is: does archaeology understand this role, does it have the resources to fulfil it?

Archaeology is not simply intended to furnish 'neutral' statements about the human past; it also has the very important purpose of helping us understand how we became what we are today — to be relevant to the contemporary world. Knowing ourselves is impossible without knowing our origins, yet neuroscientist Todd M. Preuss (2000: 1219) has referred to *Homo sapiens* as "the undiscovered primate". This relates to the perhaps greatest failure of Pleistocene archaeology, as well as paleoanthropology: these fields of research have not managed to provide us with credible information about *the human condition*. Both have provided no illumination of the circumstances and origins of our limitations, our futile yearning for everlasting life, the foundations of our constructs of the world, or our never-ending endeavours to construct meanings where there are none. Our capacity for both good and evil, the feelings and emotions associated with our existence, our 'conscious' experience of past and future, cognisance of the passage of time, and our vexed awareness of our mortality can only be explained in terms of our history as a species. No such understanding of the aetiology of the human condition has been delivered by these disciplines. But to explain human behaviour through its present-day manifestations is rather like considering illness through its symptoms instead of its causes: it explains nothing.

How should we proceed in such a daring scientific pursuit, unequalled in its significance by anything science has ever attempted? This is not an exaggeration; science has never been very effective in investigating the origins of the basis upon which human ontologies, including that of science itself (other than a superficial 'history of science'), are predicated. The question to be asked is, how effective can we expect existing knowledge and especially existing paradigms of archaeology to be in such a quest? This is what archaeology needs to be able to provide if it is to be more than a stale repository of snippets of information and haphazard explanations. It necessitates a far more broadly based analysis of archaeology than has been attempted here.

Perhaps many of the criticisms presented here are to some extent contrived and deliberate. Perhaps things are not as bad as I have described them. But even if only some of my observations were valid, we would have to concede that the discipline is in no shape to contribute to a deeper understanding of humans. It would surely involve a paradigm shift to begin moving in that direction — something not likely to happen any time soon in this troubled discipline. According to Kuhn (1993: 336), a scientific revolution occurs in part as "the transition to a new lexical structure, to a revised set of kinds". Terms for these kinds "supply the categories prerequisite to description of and generalisation about the world. If two communities differ in their conceptual vocabularies, their members will describe the world differently and make different generalisations about it" (Kuhn 1993: 319). Our language does not describe the world; it creates it (Searle 1995), and has created it since hominins first began using language (Bickerton 2010). Therefore a step in the right direction may well be to examine the taxonomy of archaeology more deliberately and meticulously.

## 'Taxonomisation' in archaeology

As shown in previous chapters, archaeology is about identity; it creates the identities of modern societies by contrasting them with 'others'. But it is also about various other things; for instance I would especially emphasise its apophenia. In statistics this refers to a Type I error or false positive: an error of excessive credulity (Allchin 2001). It is attributable to an excess in sensitivity, essentially a clustering illusion (Brugger 2001). More than anything else, archaeologists seek *patterns* in the evidence they assemble, and since their methodology is certainly governed by patterning, it is only to be expected they do find them. This is then expressed in statistical analyses of obviously non-randomly acquired data. The difference between a computed, estimated or measured value and the true or theoretically correct value is caused by non-random variations in the primary data. I have often observed this seeing of imagined patterning first hand, and it is psychologically related to an interesting phenomenon, pareidolia. The people who most keenly tell us what is depicted in rock art and what it means tend to be the same people who are the most susceptible to seeing images in random arrangements (oddly shaped flint nodules, shapes of cliffs, clouds, whatever). Indeed, the issue of susceptibility to pareidolia is one that seems to have been inadequately researched, but to explore it here would lead us too far from the course of exploring the general epistemology of archaeology. In a general sense, however, it needs to be said that the distinction between teleology (design or purpose in natural processes or occurrences) and dysteleology (purposeless in nature) is relevant here. In Pleistocene archaeology we see a subliminal tendency to regard human evolution as a teleological process, whereas in science it is assumed that the processes of nature are entirely dysteleological: there is no design, no target in evolution. In an archaeological or humanist perspective (not to mention a religious one), modern man is the pinnacle of evolution; in a scientific perspective he is a neotenous ape susceptible to the effects of an endless list of degenerative alleles, syndromes and Mendelian disorders — an evolutionary failure (Bednarik 2011a).

Archaeological excavation is a method that establishes spatial contexts of objects considered to be archaeological objects. Since archaeology, in an admittedly simplistic definition, is the study of the human past through archaeological objects, one might assume that all objects that have been made, used, touched, or in some way had a connection with humans are archaeological objects. In the sense that potentially all such articles can *become* archaeologically meaningful, this is true, but there is no intention of regarding them so until they have attracted the attention of an archaeologist. Therefore the defining quality of an 'archaeological object' is that a practitioner of the craft has taken an interest in it. This, obviously, is an entirely self-referential definition, and one that does not apply in any other discipline: an organism is what it is, irrespective of it having been studied by a biologist; even a word does not require the approval of a linguist to be a word, and chemicals exist independent of chemists.

Before this observation is rejected as flippant, we need to consider two points. First, archaeology is entirely authority based, which in science is simply unacceptable. Therefore even the pronouncement that something is or is not an archaeological object is open to challenge. Secondly, most parts of the land surface of this planet have been occupied by humans at some stage or another, over the last few million years. It follows that every place in these areas is thus an archaeological site, most of the land surface is a continuous collection of such sites, and most of the subsurface sediments are consequently archaeological deposits. That renders the demands of archaeologists that certain places need to be protected also self-referential; what they mean is that these places have attracted the attention of practitioners. Therefore this is more a matter of protecting the authority of archaeologists than it is one of protecting 'heritage': who decides what is and is not archaeological heritage? Who decides what is of sufficient archaeological importance to be protected? When we recall that cultural monuments of undeniable world significance are being destroyed by archaeologists, because such destruction can be extremely lucrative for them, the disingenuousness of this position becomes apparent. It is about authority, and authority has a prize tag attached to it. Once again, the political dimension of the discipline becomes painfully obvious.

Much of the contents of this book are likely to be rejected indignantly by many archaeologists. Theories based on different paradigms are, Kuhn says, 'incommensurable': they share no common measure. The paradigm of one period of 'normal science' may lead scientists to judge one puzzle solution more favourably than another solution, while the paradigm from another period of 'normal science' may lead scientists to make the opposite judgment. There can be no doubt that the paradigm I envisage differs considerably from that of orthodox, traditional archaeology of the kind that has evolved over the 20th century. But perhaps we can find some common ground on the issue of taxonomy. We can probably agree that the 'objective' determination of phenomenon categories would only be achievable through their CCDs (crucial common denominators, see previous chapter), and that any taxonomy derived from categories other than that can at best be provisional, or a 'working model'.

The selection of CCDs of phenomenon categories reflects the cognitive strategies of reality-building humans, but in terms of physical or objective reality it would be quite random. In simple terms, the fact that Alabama and Alaska are states of the USA is no proof that Alberta must be so too. Determining the CCD of any category is not as easy as it may seem, because so many of our percepts are misleading, especially in archaeology where there is very little to test most propositions. What renders the issue even more complicated is that, far from being a homogenous discipline, archaeology is highly compartmentalised by ideological, political, and philosophical tensions (see Chapter 2). Before we could consider the views of a practitioner, we would need to know what kind of archaeologist he or she is: a processualist, a post-processualist or a postmodernist? Perhaps a feminist archaeologist, or a Marxist, Stalinist, Biblical, fascist, socialist, behaviouralist, structuralist, middle-range, contextualist, functionalist, New Age, folk, historical, industrial, forensic, landscape, maritime, evolutionary, militarist, nationalist, colonialist, imperialist, cultural resource management, aerial, anthropological, cognitive, gender or just plain New Archaeologist? Or simply a prehistorian, ethnoarchaeologist, antiquarianist, a follower of Rathje's garbology, or any possible combination of these and other sectarian groups? Clearly these many orientations invoke very different preoccupations and perspectives of what is supposed to be neutral data. In reality, the data are not at all neutral, but are always constructed by the archaeologist through post hoc accommodative argument (Binford 1981: 31, 82–6). For instance the recognised and named stone tool types are most probably just modal points along continuums of morphological variation. Consequently the CCDs of these arbitrary types are generally unknown, and their nomenclatures consist of Searle's (1995) 'institutional facts': they lack objective validity. These or any other typologies of palaeo-archaeology depend entirely upon the selection of the common denominator variables, which we can safely assume not to be the CCDs in most instances. Consilience between these 'social sciences' and the real sciences is therefore out of the question, as the latters' epistemology shows consistently. The plausibility of 'warranting arguments' is no substitute for falsifiability, nor is 'middle range theory', and most archaeological knowledge claims are sustained by authority — in the sense that key information is not testable and must therefore be accepted on trust. Science does not trust human judgment, with good reason, and Pleistocene archaeology lacks a universal and axiomatised theoretical framework (see Chapter 6) and internal testability. As archaeologist Geoffrey A. Clark points out, for consilience to work, there must be consensus about basic definitions, terms and concepts. In my opinion, there is very little consilience in palaeo-archaeology, and almost no concern with the logic of inference underlying its knowledge claims. That said, little is to be gained by ignoring these epistemological issues. If we continue to do that, we will continue to fail to confront the fundamental ambiguity of pattern in both the archaeological and palaeontological records (Clark 2009: 30).

Clark is rightly critical of the *cultural history* approach to hominin history, advocating instead the framing of the discipline within *human behavioural ecology*. This might involve, for instance, behavioural or reproductive ecology, evolutionary psychology, dual inheritance theory, gene-culture co-evolutionary theory, niche construction theory, primatology or primate ethology, decision theory, optimal foraging theory, mating strategies and sexual division of labour, community ecology and evolutionary genetics (cf. Bednarik 2011a). Often these approaches lend themselves to hypothetico-deductive research protocols and therefore lead to testable hypotheses. Unfortunately, orthodox palaeo-archaeology has a tight grip on the discipline, through the universities, academic publishing regime, refereeing network, funding agencies, and its extensive control over public dissemination. It uses this position to reinforce its political control over the field of human origins, thus preserving the unscientific cultural history paradigm.

Which brings us to the crux of the issue. Pleistocene cultural history is derived from a series of 'cultures' that are not cultures at all, but are based on perceived combinations of tool types, usually stone tool typologies (Sackett 1981, 1988). These taxonomies present invented stone tool types rather than valid types; most of the typology derived from them is based on 'diagnostic' falsities. So since the 19th century, Pleistocene archaeology has created a vast edifice of 'cultures' through these perceived technological traditions (Petrie 1899). But even they are largely falsities: their 'tool types' are untested and untestable propositions, and may well be just fantasies of archaeologists. Moreover, knowledge of hominin history was severely constrained even a century ago, and a taxonomy proposed in the late 19th century is very probably superseded today (Bednarik 2002). This construct of hundreds of 'cultures' of pre-History is merely an elaboration of an evolutionary succession conceived before most of its details could have been evident. Constructs such as 'Bronze Age', 'Neolithic' or 'Mesolithic' (see Chapter 6 concerning the status of the latter concept) have little or no currency in most of the world (Bednarik 2002; Ziegert 2010: 28), and are in any case misleading as cultural designators. All concepts of this succession had to be compatible with a paradigm that was at best a tentative model of what the evidence *might* yield, and this becomes progressively more obvious as one proceeds back on the time scale. The concepts of a Lower, Middle and Upper Palaeolithic seem reasonable enough as provisional guides, but on close examination their definitions, however expressed, are not sustainable (Camps and Chauhan 2009). Moreover, if the definitions of the technologies concerned were applied to modern people, we would find that societies of Palaeolithic, Mesolithic, Neolithic, Chalcolithic, 'Brass Age', Iron Age and Nuclear Age technologies coexist in some countries today, e.g. in India. Which of these Eurocentric and archaeocentric pigeonholes would the Jarawas of the Andamans be assigned to? They use stone tools but they also salvage iron nails from shipwrecks and beat them into tools (Sreenatahn et al. 2008). What about Stone Age people who beat iron horseshoes into spear-points, as in the Australian Kimberley? Do the Pleistocene technological traditions of Japan defined as the Early Jomon, with their decorated ceramics, belong to the Neolithic? Where is the Bronze Age of sub-Saharan Africa, or the Mesolithic of Asia? Since these designations have no relevance outside of Europe, how can we know that they are valid even in Europe? Particularly when the concept of a Mesolithic probably derives from a misunderstanding, as already noted previously.

Each of these main phases is subdivided into a number of 'cultures', and at least in the Pleistocene their carriers are named after them. For instance one of the twenty or so recognised 'cultures' of the first half of the 'Upper Palaeolithic' is named the Aurignacian, and the people concerned are defined as the Aurignacians. But does anyone believe that there was indeed a group of people, of tribes, a nation, an ethnic entity, a language group, or any other collectively identifiable assemblage of people coinciding with this definition? Of course not — these are nonsensical words, they have no meaning outside the brains and writings of Pleistocene archaeologists. Cultures or societies cannot be defined by their utensils, although such artefacts might embody cultural information detectable to the initiated. Archaeologists have no access to such information, they simply create arbitrary taxonomies of tool types they believe they have identified.

Experts on the making and use of stone tools (e.g. traditional Aboriginal elders I have worked with in the 1960s) used taxonomies that are entirely different from those archaeologists design, and some of their CCDs would not even be detectable by the lithic typologist. This is the difference between emic information (available only to the participant of the cultural system) and etic information (consisting of 'institutional facts'). The analyst usually accepts that his classification will differ considerably from that of the user or maker of the artefacts, but maintains that his etic system is culturally neutral. This may in one sense indeed be the case, as there is at least an attempt to use objective criteria. However, nobody exists in a cultural vacuum, and in her or his 'analysis' the researcher will often need to make arbitrary decisions that will ultimately have been guided by cultural conditioning. Since these tool classes are certainly beyond falsification, being free-standing and self-confirming constructs, they cannot be counted as reliable. Yet it is the combinations of these tool types, their relative proportions within an assemblage, which then are the principal definers of the purported Pleistocene cultures.

But of even greater concern is that technologies do not define cultures, yet the variables that do so have been largely ignored in the creation of this cultural taxonomy. Since most cultural behaviour traces cannot be expected to have survived from our early history, the most comprehensive corpus of surviving cultural evidence from the Pleistocene is without question the body of palaeoart. Yet orthodox Pleistocene archaeology shows little interest in this form of evidence, and when it does it trivialises it by designating it to categories comprehensible within archaeology's simplistic reality frames of reference, i.e. by defining them as 'art objects'. This is a serious error of judgment, in that it is an application of a modern, highly specific concept to material evidence that may not be 'art' at all. Even in most ethnographically known cultures, the definition of such products as 'art' is inappropriate, so clearly here the 'analyst' uses unsuitable conceptual parameters. It is then easy to suspect that in the previous case of the stone tools, much the same may well be the case: the frame of reference the analyst applies is disadvantageous. Moreover, concerning Pleistocene paleoart, archaeological interpretations sustain a model linking the origins of these exograms (Donald 1991, 1993, 2001) to the advent of the purported African ancestors of our subspecies, another rather consequential bungle attributable to dogmatic intractability: the relevant empirical evidence to show the much earlier use of exograms has long been available (e.g. Bednarik 1992), but was either ignored or explained away.

## Summarising

These considerations provide consistent hints that the nomenclatures Pleistocene archaeologists create in their quest to find patterns in the way 'evidence presents itself' are at best unreliable, and at worst fictitious. They also show that the notion of archaeology creating a cultural history is severely mistaken. Pleistocene archaeology, in particular, is incapable of that, at least in the way it has so far operated. Therefore we need to expect that most of the constructs of the human past provided by it are likely to be falsities. In other words, as the title of this book implies, our Pleistocene history as provided by the discipline is a *creation by archaeologists*, a form of academic origins myth. But creating history is a fraught enterprise. "Who has the right to frame and interpret the past of others?" (Lyons 2002: 127; cf. Barkan and Bush 2002: 2; Brown 2003: 184). To untangle the intertwined factual and mythological parts of this contrived history will not be easy, but I have provided rudimentary tools for this task in previous chapters.

The first corrective action should perhaps be to rid us of the concept of a 'prehistory'; it is a neocolonialist notion lacking objective currency. Once it is accepted that the traditional main divisions of hominin history are in need of revision, it might be useful to replace the chronological taxonomy with such phases as incipient, archaic, developed non-sedentary, early sedentary, late sedentary, and urban societies, or some similar system that is not based on perceived tool types. But the perhaps most incisive change that needs to occur in Pleistocene archaeology and the closely allied field of paleoanthropology is a re-assessment of the purposes of these quests. Is archaeology going to continue its sterile search for evidence confirming its preconceived ideas, models and dogmas? Will the debates over minor somatic differences among former human groups continue to be fuelled by dozens more 'new species' of hominins, the main purpose of which it is that their promoters climb the academic ladder? Will interpretations continue to be framed within the ever-changing kaleidoscope of Western society's prejudices and preconceptions, or will the discipline develop the independence to operate outside of these, shaping social constructs rather than be shaped by them?

If these disciplines are only willing to provide disconnected snapshots of the human past, without an integral protocol of how and why we became what we are today, what precisely is their worth to society? The 'cultural resource management' faction of archaeology is deeply implicated in the systematic destruction of archaeological values, and the many international and national instruments supposedly regulating protection are ineffective in policing this juggernaut industry. There are good reasons for this, which I understand only too well, having worked with or against many of these agencies, and I can report that they are without exception politicised. Issues of heritage are much more closely connected to politics and tourism than to agreed intrinsic values. An example is that beautiful institution, the World Heritage List, which was hijacked by economic interests in Europe, and which even UNESCO today admits is unrepresentative and lacking in credibility.

The gatekeepers of human evolution and palaeo-ecology have completely and consistently failed in explaining why humans are the way they are. In fact some of their misguided models have been significant impediments to hard sciences, e.g. the neurosciences and cognitive sciences, retarding them needlessly in these contexts. For instance, as a result of the aggressive promotion of the replacement hypothesis by Pleistocene archaeologists in recent decades, neuroscientists had to conduct their deliberations within a false paradigm: the notion that today's humans are a distinctive species, separate from their immediate robust ancestors, whose characteristics are attributable to natural selection and genetic drift. Within that framework, the selection in favour of numerous deleterious traits has been an unsolvable paradox, as amply demonstrated by Keller and Miller (2006) and the accompanying debate. That paradox could only be explained (Bednarik 2011a) by first refuting the archaeologically derived replacement hypothesis (Bednarik 2008a) and replacing it with the domestication hypothesis (Bednarik 2008b). The stultifying infatuation of palaeo-archaeology with false interpretations has impeded the sciences on many occasions, some of which have been visited in previous chapters.

This book has shown us that archaeologists are people who tell us what happened a long time ago, making up classes of objects and entities (artefacts, motivations, intentions, social modes, practices, causal relationships, cultural groups etc.) that nobody seriously assumes to have had real existence at any time. The most appropriate motto of palaeo-archaeology would be '*ex ungue leonem*' ([to paint] 'the lion from the claw'): to deduce the whole from a small part or trace of it — which is what archaeology does at the best of times.

We have seen that archaeology is the only discipline defining its sphere of interest self-referentially. It is also the only discipline that refuses academic freedom, trying hard to restrict access to its resources. Moreover, its propositions are based on the authority of the proponent, not on testable propositions. Since excavation results in the destruction of all spatial contexts and other relevant information, we need to accept the pronouncements of the excavator (e.g. the plan and section drawings) on authority; they are not testable data. To make matters worse, excavation can only yield non-random samples, another limitation in conflict with the requirements of scientific method. The remains secured from an excavation cannot be expected to be *representative* of anything other than themselves; their composition is merely accidental, besides having been significantly affected by taphonomy. Similarly, most finds are made accidentally, and there is not much control over the historical sequence of discoveries. That the order in which they are made is largely random means that the dogma at any point in history is a similarly haphazard product, likely to be subjected to major change in the future. Therefore any endeavour to preserve the dogma is counterproductive, because the dogma must be assumed to be intrinsically false. The entire policies of the discipline conspire to keep it from attaining credibility, rather as if it had been designed to yield mostly questionable histories.

Archaeology not only destroys the sediments it excavates, it does the same with the beliefs of societies through deconstruction, objectification and academic appropriation. It is archaeologists who manage the remains and monuments of the defeated, marginalised and superseded cultures for the victorious states whose servants they are, and who license them. Throughout the history of the discipline, archaeology has helped create fictitious grandiose pasts for nation states. The political uses made of its findings have often facilitated ethnic clashes and cleansing, bigotry and nationalism (Kohl and Fawcett 1995). Acquisition of knowledge about the ontologies of indigenous societies is integral to dispossession, to diminishment of indigenous values and sovereignty, to gaining power through 'interpretation'. As we have seen in the first chapter, the American First Nations leader Vine Deloria expressed this by reminding us that Western civilisation does not link knowledge and morality but rather, it connects knowledge and power and makes them equivalent. And yet it is not even possible to translate indigenous 'metaphysics' into forms decipherable within a Western construct of reality (Berger and Luckmann 1966; Pinker 2002) without significantly corrupting it. There is again no escape from the realisation of the incommensurability of languages and social constructs.

Archaeology's ambivalences complicate its epistemological credibility further; for instance, it both supports and opposes the aspirations of indigenous peoples relating to cultural heritage. It creates taxonomies or systems of material evidence, but there is no proof that these are valid reflections of reality. It makes extensive use of the sciences and seems to have aspirations of becoming a science, yet it maintains a non-scientific epistemology by rejecting principles of falsifiability and blind testing. Archaeology values its material evidence and jealously guards it, demanding a monopoly, yet it is also the most effective destroyer of this evidence and its prohibitions contradict the principles of academic freedom. Archaeology destroys most of its evidence — not always intentionally, but because it lacks the methods and understanding it has yet to gain. At other times, archaeology destroys archaeological materials very deliberately (Bednarik 2006, 2008c).

All classifications archaeology produces are *arbitrary constructs of specialists.* These classes of objects, entities and interpretations are etic pronouncements without real existence. They are fantasies created through autosuggestion, 'observer-relative or institutional facts' (see Chapter 7). Every occupation layer is called a cultural horizon, even though it rarely yields any truly cultural information (in the correct sense of that term). Purported tool types are recruited to form technocomplexes, and these are then defined as cultures, even though tools are certainly inter-cultural entities and are unsuitable for defining cultures. Authentic cultural variables, such as palaeoart, are ignored in the definition of Pleistocene and early Holocene cultures. Even the taxonomies of these tool types are etic and very probably false, yet they and the combinations in which they occur were used to invent all named cultures. To make matters worse, these mythological cultures are then attributed to specific groups of people who are named after them.

We have also seen that archaeology is based on fetishes: objects are regarded as representing people, and if specific artefact types (beakers, for instance) are perceived to spread geographically over time, it is taken as proof of the mass movement or migration of peoples. Similarly, if genes are thought to have spread spatially, this proves for archaeologists that entire populations moved and replaced others. In this fetishism it is ignored that there are far more effective, and far more likely explanations for the movement of both 'memes' and genes. Reticulate introgression, in the latter case, accounts for such travel of genes in the natural world, but archaeologists prefer the 'wandering tribes' model, a model lacking biological justification. But even if authentic cultural variables were used to define cultures, it would be of limited help, because often archaeologists cannot agree on which cultural evidence (e.g. cave art) belongs to which perceived tool industry (e.g. Clottes et al. 1995; Zuechner 1996; Pettitt and Bahn 2003; Valladas and Clottes 2003; Valladas et al. 2004, concerning just Chauvet Cave). Whichever way one turns, the paradigms of archaeology are haunted by the incommensurabilities of its empirical components.

Ultimately it seems that the notion of archaeology begins with an idealistic search for knowledge about the human past, but that its 'professionalisation' has destroyed the innocence of that idea. The demands of instititionalisation and careerism are among the reasons for the fragmentation and vagueness of the discipline as defined in Chapter 2, because as each ambitious career archaeologist tries to make his or her mark, new directions are created and then zealously promoted. Another reason, obviously, is that the establishment of such diffuse niche interests is facilitated by a field that has no underlying universal theory and little rigour, in which 'anything goes' — as long as it is done by vocational (i.e. paid) practitioners. Avocational archaeologists, on the other hand, are fiercely opposed by Campbell's (2006) "molesters of the past": only 'professionals' should be permitted to molest the past. But the most spirited opposition is reserved for those 'amateurs' whose archaeological knowledge matches or — worse still — exceeds that of most dependent archaeologists. And this is perhaps the most nefarious aspect of state-run archaeology, because obviously it is not the competition for positions or public funds that is feared. So what is it that generates such odiousness in some dependent practitioners when dealing with erudite independent colleagues?

Again, the answer is probably to be found in the inadequate epistemology of the discipline, fostering a latent insecurity and a fear of 'outsiders'. In this book I have tried to allay this anxiety by showing that there needs to be no further torment concealing the weaknesses of archaeology; they have been identified, they are quite well understood, therefore further apprehension of this kind is unnecessary. This book has shown us that archaeologists no longer need to hide the fact that they have failed to establish what happened in the human past; or that they cannot find agreement on any aspect of human history; or even on what precisely archaeology is. This realisation from the present volume will hopefully remove the deep suspicion many dependent archaeologists have of knowledgeable independent ones, and perhaps it can even lead to treating them as valued colleagues. In Frank Campbell's words, as quoted in Chapter 1, "[t]he tragedy is that archaeology has promised a grand narrative but can deliver only conjecture. The archaeologist has no clothes." This book has shown that Campbell is right. Hence there is nothing more to conceal or to fear; 'professional' archaeologists can 'come out' and admit what we already know.

In the final analysis, archaeology is a hobby that somehow got a little out of hand. It could revert to being just a hobby, or alternatively it could change direction and become a science. But there is nothing unusual or unique in this situation: it pertains to the other humanities as well (Bednarik 2011b).

## REFERENCES

Allchin, D. 2001. Error types. *Perspectives on Science* 9(1): 38–58.

Barkan, E. and R. Bush 2002. Introduction. In E. Barkan and R. Bush (eds), *Claiming the stones, naming the bones: cultural property and the negotiation of national and ethnic identity*, pp. 1–13. Getty Research Institute, Los Angeles.

Bednarik, R. G. 1992. Palaeoart and archaeological myths. *Cambridge Archaeological Journal* 2(1): 27–43.

Bednarik, R. G. 2002. The human ascent: a critical review. *Anthropologie* 40(2): 101–105.

Bednarik, R. G. 2006. *Australian Apocalypse. The story of Australia's greatest cultural monument.* Occasional AURA Publication 14, Australian Rock Art Research Association, Inc., Melbourne.

Bednarik, R. G. 2008a. The mythical Moderns. *Journal of World Prehistory* 21(2): 85–102.

Bednarik, R. G. 2008b. The domestication of humans. *Anthropologie* 46(1): 1–17.

Bednarik, R. G. 2008c. More on rock art removal. *South African Archaeological Bulletin* 63(187): 82–84.

Bednarik, R. G. 2011a. *The human condition*. Springer, New York.

Bednarik, R. G. 2011b. Rendering humanities sustainable. *Humanities* 1(1): 64-71; *http://www.mdpi.com/2076-0787/1/1/64/*

Berger, P. L. and T. Luckmann 1966. *The social construction of reality: a treatise in the sociology of knowledge*. Anchor Books, Garden City, NY.

Bickerton, D. 2010. *Adam's tongue: how humans made language, how language made humans*. Hill and Wang, New York.

Binford, L. R. 1981. *Bones: ancient men and modern myths*. Academic Press, New York.

Brown, M. F. 2003. *Who owns native culture?* Harvard University Press, Cambridge, Mass.

Brugger, P. 2001. From haunted brain to haunted science: a cognitive neuroscience view of paranormal and pseudoscientific thought. In J. Houran and R. Lange (eds), *Hauntings and poltergeists: multidisciplinary perspectives*. McFarland & Company, North Carolina.

Campbell, F. 2006. Molesting the past. *The Weekend Australian*, 25 February: R15.

Camps M. and P. R. Chauhan (eds) 2009. *Sourcebook of Paleolithic transitions: methods, theories, and interpretations*. Springer, New York.

Clark, G. A. 2009. Accidents of history: conceptual frameworks in paleoarchaeology. In M. Camps and P. R. Chauhan (eds), *Sourcebook of Paleolithic transitions: methods, theories, and interpretations*, pp. 19–41. Springer, New York.

Clottes J., J.-M. Chauvet, E. Brunel-Deschamps, C. Hillaire, J.-P. Daugas, M. Arnold, H. Cachier, J. Evin, P. Fortin, C. Oberlin, N. Tisnerat and H. Valladas 1995. Les peintures paléolithiques de la Grotte Chauvet-Pont d'Arc, à Vallon-Pont-d'Arc (Ardèche, France): datations directes et indirectes par la méthode du radiocarbone. *Comptes Rendus de l'Académie des Sciences de Paris* 320, Ser. II: 1133–1140.

Donald, M. 1991. *Origins of the modern mind: three stages in the evolution of culture and cognition*. Harvard University Press, Cambridge, MA.

*Donald, M. 1993. On the evolution of representational capacities. Behavioural and Brain Sciences 16: 775–785.*

*Donald, M. 2001. A mind so rare: the evolution of human consciousness.* W. W. Norton, New York.

Keller, M. C. and G. Miller 2006. Resolving the paradox of common, harmful, heritable mental disorders: which evolutionary genetic models work best? *Behavioral and Brain Sciences* 29: 385–452.

Kohl, P. L. and C. Fawcett (eds) 1995. *Nationalism, politics, and the practice of archaeology*. Cambridge University Press, Cambridge.

Kuhn, T. S. 1962. *The structure of scientific revolutions*. University of Chicago Press, Chicago.

Kuhn, T. S. 1993. Afterwords. In P. Horwich (ed.), *World changes. Thomas Kuhn and the nature of science*, pp. 311–341. MIT Press, Cambridge, Mass.

Laland, K., J. Odling-Smee and M. W. Feldman 2000. Niche construction, biological evolution and cultural change. *Behavioral and Brain Sciences* 23: 131–175.

Leone, M. and P. Potter 1992. Legitimation and the classification of archaeological sites. *American Antiquity* 57: 137–145.

Lyons, C. L. 2002. Objects and identities: claiming and reclaiming the past. In E. Barkan and R. Bush (eds), *Claiming the stones, naming the bones: cultural property and the negotiation of national and ethnic identity*, pp. 116–137. Getty Research Institute, Los Angeles.

Odling-Smee, F. J., K. N. Laland and M. W. Feldman 2003. *Niche construction: the neglected process in evolution*. Princeton University Press, Princeton, USA.

Petrie, F. 1899. Sequences in prehistoric remains. *The Journal of the Anthropological Institute of Great Britain and Ireland* 29(3/4): 295–301.

Pettitt, P. and P. Bahn 2003. Current problems in dating Palaeolithic cave art: Candamo and Chauvet. *Antiquity* 77: 134–141.

Pinker, S. 2002. *The blank slate. The modern denial of human nature*. Penguin Putnam.

Popper, K. 1934. *Logik der Forschung*. Julius Springer Verlag, Vienna.

Popper, K. 1970. Normal science and its dangers. In I. Lakatos and A. Musgrave (eds), *Criticism and the growth of knowledge*, pp. 51–58. Cambridge University Press, Cambridge.

Preuss, T. M. 2000. What's human about the human brain. In: M. S. Gazzaniga (ed.), The new cognitive neurosciences, pp. 1219–1234. MIT Press, Cambridgc, MA.

Sackett, J. R. 1981. From de Mortillet to Bordes: a century of French Paleolithic research. In G. Daniel (ed.), *Towards a history of archaeology*, pp. 85–99. Thames and Hudson, London.

Sackett, J. R. 1988. The Mousterian and its aftermath: a view from the Upper Paleolithic. In H. L. Dibble and A. Montet-White (eds), *The Upper Pleistocene prehistory of western Eurasia*, pp. 413–426. University of Pennsylvania, Philadelphia, PA.

Searle, J. R. 1995. *The construction of social reality*. Allen Lane, London.

Sreenathan, M., V. R. Rao and R. G. Bednarik 2008. Palaeolithic cognitive inheritance in aesthetic behavior of the Jarawas of the Andaman Islands. *Anthropos* 103: 367–392.

Valladas, H. and J. Clottes 2003. Style, Chauvet and radiocarbon. *Antiquity* 77: 142–145.

Valladas, H., J. Clottes and J.-M. Geneste 2004. Chauvet, la grotte ornée la mieux datée du monde. *À l'Échelle du Millier d'Années* 42: 82–87.

Ziegert, H. 2010. *Adam kam aus Afrika — aber wie? Zur frühesten Geschichte der Menschheit.* Universität Hamburg, Hamburg.

Zuechner, C. 1996. The Chauvet Cave: radiocarbon versus archaeology. *International Newsletter of Rock Art* 13: 25–27.

# INDEX

## A

B



C

## D

E



F

## I



## J



## K

## P

## Q



## R



## S

## T

U



V



W



Z